A FIELD GUIDE TO
LONG ISLAND'S SEASHORE

Glenn A. Richard by

GLENN A. RICHARD
LINDA SPRINGER-RUSHIA
PAMELA G. STEWART

Pamela G. Stewart

Illustrations by Maria T. Weisenberg

Museum of Long Island Natural Sciences
Stony Brook, New York

A Field Guide to Long Island's Seashore

Published by
Museum of Long Island Natural Sciences
State University of New York at Stony Brook
Stony Brook, New York 11794-2100
Telephone: 631-632-8230
www.molins.sunysb.edu/
© 2001 by Museum of Long Island Natural Sciences
All rights reserved. Published 2001

Library of Congress Control Number: 2001091116

ISBN 1-892170-06-X

STATE UNIVERSITY OF NEW YORK

Acknowledgments

This sequel to our first two publications, *A Field Guide to Long Island's Woodlands*, and *A Field Guide to Long Island's Freshwater Wetlands*, has been made possible by the efforts of the following individuals:

Ann Lattimore of the Mineral Physics Institute, Center for High Pressure Research, for contributing her time and talent at desktop publishing; **Jeanne Perry**, for the compilation of the glossary and index, along with editing, and great restaurant recommendations; and **Jean Cole** for proofreading and her sense of humor.

A very special thank you goes to the **Paul Simon Foundation** for their generous support of this project. Funding was also provided by the **Center for High Pressure Research**, a National Science Foundation Science and Technology Center.

TABLE OF CONTENTS

INTRODUCTION

Beaches - the very word evokes thoughts of sunbathers on a hot summer day, children sculpting sandcastles, couples strolling hand-in-hand along the shore, sailboats bobbing in the bays, and fiery, red sunsets over an oceanic horizon. From the Rockaways to Montauk, Port Washington to Orient, Long Island's hundreds of miles of shoreline are home to many magnificent beaches. A few, such as Jones Beach State Park, have been voted "best beaches in the country".

Nothing compares to the exhilarating smell of salt spray in the air and the feel of walking on the sand with wind whipping your face and hair. Local seashores offer year-round natural beauty such as dune grasses waving in the breeze, seagulls soaring and swooping, colorful windrows of seashells washed upon the sand and tidal pools teeming with sea stars. Seasonal splendor brings winter snow swirling on the sand, the strident call of the Red-wing Blackbirds heralding spring, summer's bounty of purple Beach Plums, the scarlet-stemmed Saltwort and fresh bay scallops of autumn. The coastlands provide a plethora of opportunities and experiences for casual beachcombers as well as for the more advanced bird-watchers, shell collecters and other naturalists.

The beach has an air of mystery about it. Have you ever wondered why some beaches are rocky while others are sandy? Do Horseshoe Crabs sting? What are those black, rectangular, four-pointed objects washed up on the shore? How does a starfish eat? Where does a hermit crab get its shell? This book addresses these and many other questions pertaining to Long Island's seashore. Take it to the beach and have fun!

How To Use This Book

This book is intended as an introduction to the seashore of Long Island. The types of coastline covered here include the salt marsh, estuaries, bays, lagoons, sound, ocean, and beaches. Topics such as geology, geologic history, ecology and human interaction will be discussed in the opening chapters.

As in its companion books, *A Field Guide to Long Island's Woodlands* and *A Field Guide to Long Island's Freshwater Wetlands,* the plant species are grouped by growth habit or other morphological distinctions. Trees, shrubs and wildflowers, aquatic marine plants, grasses, rushes, and seaweeds have their own sections. Animals are grouped taxonomically, with sections devoted to most of the vertebrate classes and additional sections to the invertebrates. Common and scientific names are provided for each species, along with pertinent characteristics. Habitats are specified to assist you in finding the seashore biota and include community, substrate and depth, where applicable. Illustrations of most species are provided for your enjoyment and to aid you in their identification. To clarify scientific terminology and to locate a particular species, a glossary and an index have been included. This book is not intended as a complete monograph, but instead presents a sampling of what you typically can encounter at Long Island's coastline and in its surrounding waters.

CHAPTER I

Geological Setting of Long Island and its Coastline

Geologic Features of Long Island

Long Island is adjacent to the Atlantic Bight, a wide embayment that stretches from Cape Cod, Massachusetts to Cape Hatteras, North Carolina. Since the Island is situated near the northern limit of this body of water, where the coast is oriented in an east by northeasterly direction, it faces a huge expanse of the Atlantic to its south. As a result, northeasters and hurricanes frequently strike Long Island after they become energized by abundant moisture during their long journey across the water.

The surface of Long Island is covered by a thick layer of sediment deposited by a series of continental ice sheets that originated in Canada and expanded during times of cooler climates. During these periods, more snow accumulated during the winter than was able to melt during the summer. The ice sheets, or glaciers, flowed out in all directions under their own weight, scraping up sediment, plucking bedrock from underneath themselves and carrying this material with them. Some of the material was directly deposited by ice as glacial till, a mixture of sediment of a wide range of sizes, from tiny clay particles to boulders.

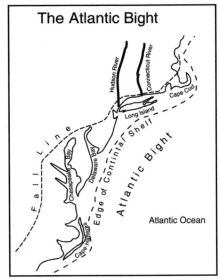

The Atlantic Bight

Hudson River

Connecticut River

Cape Cod

Long Island

Fall Line

Chesapeake Bay

Delaware Bay

Edge of Continental Shelf

Atlantic Bight

Cape Hatteras

Atlantic Ocean

The till lies within lines of hills called moraines with an east-west orientation. The Harbor Hill Moraine runs along Long Island's North Shore, including the North Fork, while the Ronkonkoma Moraine runs through central Long Island and the South Fork. Additional smaller moraines also occur, such as the Peconic Bay Moraine that lies, in part, on Shelter, Robins, and Gardiners Islands. The moraines mark various positions where the edges of ice sheets were located. Meltwaters from the ice sheets deposited sediment called outwash in a series of broad coalescing deltas that formed flat areas known as outwash plains south of the moraines. Two major outwash plains are the Hempstead Outwash Plain along the southern part of Long Island, extending into the Atlantic Ocean, and the Terryville Outwash Plain, between the Ronkonkoma and Harbor Hill Moraines. On Long Island, the glacial outwash is mainly sand with some gravel. Outwash occurs in the moraines as well as in the outwash plains.

At times when the sediment dried out, some of the silt and finer sand was picked up by the wind and deposited as loess. After the last ice retreat, over a period of thousands of years, wind formed a field of dunes at Friars Head on eastern Long Island. Coastal erosion later truncated the northern part of this dune field, forming a spectacular series of coastal bluffs. More recently, development destroyed a major portion of the remaining dunes.

Lakes sometimes formed between the edge of the ice sheets and other barriers such as the moraines. Thick deposits of lacustrine clay, such as the Smithtown Clay, over 100 feet thick in Nissequogue, formed in these lakes. Lacustrine strata, till, outwash, and loess are exposed in numerous

4

coastal bluffs along Long Island's North Shore as well as the South Fork and on the islands in Peconic Bay. In a few locations, for example Caumsett State Park and Garvies Point, much older sediment is exposed in coastal bluffs. This material was deposited in huge river deltas while dinosaurs still existed during the Cretaceous Period.

Sea Level Rise

Sea level has risen markedly since the last ice maximum, which occurred about 21,000 years ago. When the ice was at its greatest extent, sea level at Fire Island was 130 meters below what it is at present. At that time, the coast was miles offshore of its current position. The Hudson River eroded a deep canyon into the soft coastal plain sediments that were exposed by the low sea level. Today, the Hudson Canyon is submerged beneath the waters of the Atlantic due to sea level rise. Most of this rise has been brought about by the water that was added to the oceans by the melting of the ice. However, some of the recent rise in the area of Long Island also seems to be the result of ongoing subsidence of the underlying crust along a portion of the East Coast of the United States. Since sea level rise began over 21,000 years ago, its rate has varied, but over the last few centuries, it has proceeded at an average of 3 millimeters per year.

What's in Seawater?

Seawater is composed mostly of water, but unlike freshwater, contains considerable amounts of dissolved solids. A measure of the quantity of dissolved solids is termed salinity. It is usually expressed in terms of grams of dissolved solids per kilogram of seawater, or parts per thousand.

In the open Atlantic, the salinity is about 36 parts per thousand, but it varies markedly with location. Near the coast, freshwater mixes with the seawater, especially near the mouths of streams and at inlets, and this mixing dilutes the dissolved solids, lowering salinity. Areas where a significant amount of this mixing takes place are called estuaries. Long Island Sound and Peconic Bay are examples of estuaries. In central Long Island Sound, the salinity is about 30 parts per thousand, and near the mouth of the Nissequogue River, it is even lower and varies with the tides and weather. When saltwater becomes stranded in pools during low tide on warm, sunny days, some of the water may evaporate, resulting in a higher concentration of dissolved solids. This may occur in certain areas within salt marshes where it can bring the salinity up to 40 parts per thousand or higher.

If one evaporates all the water from a sample of seawater, the dissolved solids are left behind as salt. This remaining precipitate has a composition that differs somewhat from that of table salt. By mass, sea salt consists of 55.29% chloride, 30.75% sodium, 7.74% sulfate, 3.69% magnesium, 1.17% calcium, 1.14% potassium and smaller amounts of numerous other substances, including all elements that are necessary to sustain life. Although salinity varies within the sea, the proportion of elements within the dissolved solids remains about the same in nearly all marine environments. Exceptions occur around some deep sea vents, where water from the Earth's interior enters the sea.

As salinity increases, the freezing point and temperature of maximum density of seawater decrease. Freshwater freezes at 0° Celsius and its temperature of maximum density is about 4° Celsius. Therefore, when a freshwater lake cools in the winter, the water becomes denser and sinks as it cools at the surface. After the entire lake has reached the temperature of maximum density of water, further cooling decreases the density, and the frigid water tends to float at the surface. Finally, the water on top reaches the freezing point and becomes ice, which is of even lower density, and floats at the surface. This ice insulates the underlying liquid water from the wind, and circulation becomes nil. The ice gradually thickens below the surface, but except in shallow ponds, the bottom will likely remain unfrozen throughout the winter.

At a salinity of 24.7 parts per thousand, the temperatures at which freezing and maximum density of water occur are identical at −1.33° Celsius. As a consequence, in bays and other restricted areas with seawater, the water will sink to the bottom as it cools until the entire body of water is just about at the freezing point. If the air stays cold and the water is shallow, the water may quickly freeze all the way down to the bottom. In salt marshes on Long Island, a thick layer of ice often forms that rises and falls with the tide.

Waves

Waves are rhythmic movements of water that are generated by disturbances such as wind, earthquakes, or gravitational forces. The familiar waves that we see breaking along the shore are the wind generated ones. Waves from earthquakes, called tsunamis, are very long and low while out at sea. As they reach the shoreline, they steepen tremendously and crash upon the shore, causing a huge amount of destruction. Tsunamis are rare and extremely unlikely to occur on Long Island. Waves caused by gravitational action of the moon and sun are the tides.

A series of wind generated waves appears as a set of long, parallel, elevated crests separated by low troughs. In wind generated waves out on the open ocean, the individual water molecules participate by moving in circular or elliptical patterns. This causes the rise and fall of the water as the crests and troughs advance in a direction perpendicular to their lengths. Since water molecules move in cyclical patterns, they do not

Anatomy of a Wave

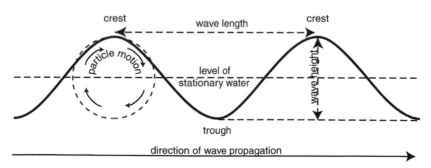

advance along with the wave. What is ultimately propagated by a wave is energy of motion. Waves pick up energy from the wind and carry it to the shore.

The size of a wave is determined by the velocity of the wind, the duration with which it blows and the distance over which it is in contact with the water. The fierce winds of hurricanes and other storms result in huge waves. During northeasters, high winds may be sustained for several days, also bringing about large waves. On the Atlantic Ocean, the distance over which the wind can generate and build waves, called the fetch, is greater than on the Sound and Peconic Bay, so Long Island's largest waves are from the ocean along the South Shore. In the open Atlantic, the height of waves, or vertical distance from trough to crest, may be as much as 75 feet during large hurricanes.

The wavelength is the distance between successive crests or troughs. As the wave reaches shallow water where the depth is about half the wavelength, interaction of the cyclical movement of the water with the bottom slows the advance of the wave. The steepness increases, and when the wave height reaches about three quarters of the water depth, it becomes unstable and breaks. The crashing of the wave stirs up the underlying sediment into irregular, turbulent motions. As a wave breaks, bursting bubbles form salt spray, a vapor of water and sea salt that permeates the air and has important effects on plants and human-made materials. Some of the salt spray becomes deposited on the dunes and swales and is a major contributor of nutrients needed by plants.

Wind-generated waves often approach the shoreline at an angle, result-ing in a wave front that is not quite parallel to the shore. Consequently, the part of the wave that is closest to shore reaches shallow water before the remainder of it does, and this slows down that portion of the wave. The result is that on a straight shoreline, waves that are oblique to the beach tend to become more nearly parallel to it as they reach the shore and break. This phenomenon is known as wave refraction. However, they still strike the shore at an angle, and this results in a process called beach drift.

When the wave washes up on the shore, the uprush, or swash, carries particles of sand in the direction that it is moving. The oblique direction results in particles being carried up the beach at an angle. As the water rushes back down the beach, the backwash carries particles nearly straight toward the ocean, but the water loses energy as some of it sinks into the sand, leaving behind a portion of the sediment. In this manner, beach drift moves sand along the shoreline.

Strong winds may sometimes pile up water against the shore. This water flows along the shore in the same direction as the beach drift and is called longshore drift. Like beach drift, this process can carry sediment parallel to the shore.

Beach drift and longshore drift are collectively known as littoral drift. Along the South Shore, the drift is generally in an east to west direction, but on the North Shore, due to its irregularity, the situation is much more complex. As exemplified by the northern extremity of Eaton's Neck, beach drift typically occurs in different directions on opposite sides of a prominent headland. Nissequogue is known as a winged headland because it has one spit growing toward the west adjacent to the mouth of the Nissequogue River and another spit accreting toward the east by Stony Brook Harbor.

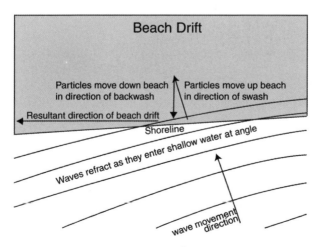

When beach drift carries sediment toward open water, deposition may extend the beach into the open water forming a spit. Examples of spits on Long Island include the western ends of Fire Island and the Rockaway Peninsula, Long and Short Beaches in Nissequogue and Orient Beach State Park near the eastern end of the North Fork.

Along shorelines composed of adjacent headlands and embayments, such as Long Island's North Shore, the water is usually shallower seaward of the headlands. Consequently, the parts of waves approaching the headlands tend to slow down and become refracted toward headlands. The result is that the energy of the waves is concentrated on the headlands, while less of the energy reaches the bays. Due to the high wave energy on the headlands, particles tend to be eroded from them and settle in the bays where the energy is lower. By removing material from the headlands and filling in the neighboring bays, waves tend to straighten out irregular shorelines.

Tides

Tides are caused by the gravitational force of the moon and the sun. Although the sun's pull is much stronger than that of the moon, due to its greater mass, the moon is sufficiently closer to our planet for it to exert a more powerful force on the Earth.

The gravitational forces of the sun and moon are greatest on the part of the Earth that is closest to them at a given time. As a result, the water particles on a part of the Earth that faces the moon are pulled toward the moon to a greater degree than the Earth itself. This creates a tidal bulge. In addition, the Earth is pulled toward the moon more strongly than the water that lies on the side facing away from the moon, causing another bulge on that side.

During new and full moons, which occur alternately about every two weeks, the Earth is in alignment with the sun and moon, which brings about a maximum tidal range. These are called spring tides. When the moon is in its first or third quarter and the side facing the Earth is half illuminated, the moon, Earth, and sun form a right angle, and the range of tides is at a minimum. These are called neap tides.

Since water must flow from one place to another for tides to occur, the tidal range is strongly influenced by the morphology of the coastline. Landforms that bound bays, sounds, harbors, and lagoons restrict the flow of water, reducing the tidal range. However, landforms can also have a funneling effect, which increases the tidal range at the narrow landward end of a body of water. This occurs in Long Island Sound, which has a tidal range of only two feet at the eastern end that is open to the ocean, but experiences tides of six feet at the shoreline of western Nassau County.

Shallow restricted inlets can also bring about an asymmetry in the tidal cycle. In the open ocean, the tides are free to rise and fall as the Earth turns and its orientation changes relative to the moon and sun. A marsh, such as Flax Pond, which is connected to Long Island Sound by a narrow and shallow inlet, approaches high tide soon after high tide is reached in the Sound. However, as the tide falls, the amount of water in the inlet drops, increasingly restricting flow as low tide is approached in the Sound. When low tide finally occurs in Long Island Sound, the water level in Flax Pond is still falling because the restricted inlet has not enabled the marsh to empty quickly enough to keep pace with Long Island Sound. It is not until two hours after low tide has occurred in Long Island Sound, where the water is rising again, that the water in Flax Pond manages to drop to the level of the Sound. Therefore, the ebb tide at Flax Pond is approximately eight hours in duration, while the flood tide runs for only about four hours.

The Gulf Stream

The Gulf Stream is a warm current that originates in the Gulf of Mexico and heads northeast, parallel to the Atlantic coast of the United States, gradually cooling and mixing with the surrounding waters until it loses its identity well off the coast of New England. It moderates the temperatures of the nearby coastal areas, including Long Island, keeping them warmer in winter than many other areas of similar latitude. The current often brings organisms from southern waters to Long Island's coast, especially along the South Shore.

Beaches

A beach is considered to be an accumulation of material, such as sand, gravel, other sediment and shell material that is moved and deposited along the shore by waves. The part of the beach farthest from the shoreline is the back beach, or berm, which is composed of material deposited there by large waves during storms or the highest tides. This area is usually rather flat and may even slope away from the shoreline. Seaward of this feature is the beach face which is the area that lies between low and high tides and slopes toward the water. This area is constantly washed by waves. Below the beach face is the shoreface, which is always submerged but shallow enough to be affected by the action of the waves on the bottom sediments. On the ocean shoreline, it extends out to where the water is about 30 feet deep. On Long Island Sound, where the waves are smaller, the lower limit of the shoreface is shallower.

Barrier Beaches and Spits

Barrier beaches line much of the East and Gulf Coasts of the United States south of Cape Cod. Most of the South Shore of Long Island is composed of barrier beaches. These beaches are long and narrow accumulations composed mostly of sand that is deposited by waves and currents. The sand is reworked continually by wind and water.

Fire Island is Long Island's largest example of a barrier beach. It is an island bound by the open Atlantic Ocean to the south, Great South Bay, an example of a lagoon, to the north, and an inlet at each end that connects the lagoon to the Atlantic. Fire Island contains dunes which are composed of sand that was carried from the beach by wind. Tides move in and out of the lagoon through the inlets, with the currents moving sediment. As the water moves past the inlet in either direction, it fans out and drops some of this load. The result is a flood tide delta on the lagoon side and an ebb tide delta on the ocean side. Ebb tide deltas are

Cross Section of the
Fire Island Barrier Beach

Vertical scale is exagerrated

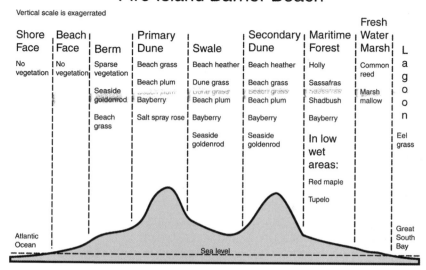

Shore Face	Beach Face	Berm	Primary Dune	Swale	Secondary Dune	Maritime Forest	Fresh Water Marsh	Lagoon
No vegetation	No vegetation	Sparse vegetation	Beach grass	Beach heather	Beach heather	Holly	Common reed	
			Beach plum	Dune grass	Beach grass	Sassafras		
		Seaside goldenrod	Bayberry	Beach plum	Beach plum	Shadbush	Marsh mallow	
		Beach grass	Salt spray rose	Bayberry	Bayberry	Bayberry		
				Seaside goldenrod	Seaside goldenrod	In low wet areas:		Eel grass
						Red maple		
						Tupelo		
Atlantic Ocean							Great South Bay	

Sea level

subject to wave action which helps keep particles suspended and reworks much of the material that does become deposited. Therefore, they are generally not as well developed as flood tide deltas.

In 1858, a lighthouse was built at the western end of Fire Island. Since that time, the Island has grown steadily toward the west and now the lighthouse is five miles from the end.

A number of theories have been proposed for the origin of barrier beaches. Most have probably originated as spits that were lengthened by littoral transport, then breached by the formation of inlets. Subsequently, the ends of the spits were cut off to become islands. Another hypothesis is that a succession of beach ridges formed adjacent to the coast as sea level rose during the last ice retreat. As sea level continued to rise, the low areas behind the beach ridges were drowned and became the lagoons that we now find behind the barrier islands. The beach ridges themselves became the barrier islands. A third hypothesis is that the emergence of offshore bars as they were built up by waves formed barrier islands.

Rampino and Sanders (1981) found that about 9000 years ago, a series of barrier islands formed 7 kilometers south of the present coastline. This was when the sea was 24 meters lower than it is now. 7000 years ago, these islands were drowned as sea level rose rapidly. By about 6000 years ago, a new chain had formed about 2 kilometers south of the present shore and it has gradually migrated northward since that time.

In any case, there are two main sources for the sand that makes up the barrier beach system along the South Shore of Long Island. Sediment is eroded by wave action from the headlands of the South Fork and distributed on the adjacent beaches. Littoral transport then gradually moves this sediment toward the west. The glacial outwash sediments on the continental shelf to the south of Long Island are another source. In fact, an enormous amount of sand lies offshore, and this may help to stabilize the position of Fire Island and other barrier beaches, at least temporarily, as sea level rises.

A walk from the ocean to the lagoon reveals the types of features that make up a barrier beach. The area next to the ocean that is regularly washed by waves is the shoreface. It slopes toward the sea. An accumulation of sand behind the shoreface is the berm. In back of the berm may be a line of primary dunes composed of sand deposited by wind. Currently, a low area or swale may lie behind the primary dunes, with secondary dunes behind the swale.

During large storms, waves may wash through the dunes and the water may flow to the lagoon. This can cause a great deal of erosion and may even form a new inlet. This process, called overwash, can carry sediment from the beach to the lagoon, building up the landward side of the barrier island as the seaward side erodes. However, erosion and deposition on the seaward side are influenced by seasonal and sporadic events, so trends can be difficult to discern without long term observation.

Mineralogy of Fire Island Sand

Sand on Fire Island is 98% quartz, with the remainder being almost entirely magnetite, garnet and tourmaline. These minerals, while mixed together in most places, are not evenly distributed on the beach.

Quartz (SiO_2) is usually a white or clear mineral with a specific gravity of 2.65. This means that its density is 2.65 times that of pure water. It is abundant on Long Island due to the fact that it is relatively insoluble in water, resistant to chemical weathering and important as a component of local rocks. In contrast, feldspar, which is also a significant mineral component of these rocks, including cobbles and boulders on beaches, is almost nonexistent in the sand on Fire Island. This is because it weathers rapidly in the presence of moisture.

The almandite garnet ($Fe_3Al_2(SiO_4)_3$) that occurs on Fire Island is a reddish to purple mineral with a specific gravity of 4.25, meaning that it is considerably denser than quartz. Magnetite (Fe_3O_4), with a specific gravity of 5.18 is black and magnetic. Tourmaline, which has a complex and variable chemical formula, is also black, but not magnetic, and has a specific gravity of 3.0 to 3.25. Garnet, magnetite, and tourmaline are all resistant to chemical weathering. Their lower abundance in beach sand is accounted for by their lesser occurence than quartz in rocks.

Given an equal grain size, garnet, magnetite and tourmaline require higher wind or water velocities to be picked up than quartz due to their higher density. Therefore, wind or water moving at certain speeds will be able to pick up the quartz while leaving these heavier minerals behind. At some locations on the beach, the quartz is removed, resulting in a mixture of these darker minerals known as lag deposits. On sunny days, these absorb more heat from the sun than lighter sand and may become quite hot.

The minerals that make up beach sand are very low in plant nutrients. On Fire Island, most of the nutrients in the sand are from salt spray.

Dunes

Dunes are accumulations of sand deposited by wind. Waves supply the sand by building up the beach to the elevation of the highest tides. These tide levels are realized during full and new moons and storms. Since most tides do not reach this high, the material deposited under these conditions is beyond the limit of most waves.

After significant portions of a spit or a barrier island have been built up above the reach of most waves, the sand tends to dry out and is subject to being lifted and carried by wind. When this occurs, dunes may begin their existence as small accumulations of sand behind a piece of driftwood or another object lying on the beach. The object creates a wind shadow which causes sand that was picked up by the wind to drop to the ground. This happens because wind, which has reached a high enough velocity to carry sand, is slowed down when it reaches the wind shadow and loses the energy that enabled it to pick up the sand. Once the sand begins to accumulate, it forms its own wind shadow and more sand can accumulate behind it, building up the nascent dune.

The dune's increasing elevation above the reach of waves also provides an opportunity for vegetation to grow more densely, and the stems and leaves of these plants help trap even more sediment. The roots bind the sediment, offering some resistance to erosion.

Dunes are characterized by steep slopes which represent the angle of repose of the material of which they are composed. This is the maximum angle at which an accumulation of the material is stable. Since the surfaces of dunes are beyond the reach of most waves, vegetation is able to become established there. The roots and rhizomes of the vegetation provide some stability for the dunes, as they tend to resist erosion. However, storm waves, most often in the winter, may wash over the berm and attack the shoreward face of the dunes. This may erode some of the sand and leave a vertical scarp several feet high. Its slope may be ninety degrees, which greatly exceeds the angle of repose. An inspection of this shoreward face reveals that it contains roots and rhizomes that became exposed by the erosion. This is what enables the sand to lie at an angle larger than its angle of repose.

Dunes on Atlantic Ocean beaches are often struck by waves during large storms. This may erode the face of the dune, exposing roots and layering inside. The layering effect consists of light and dark deposits of sand. The light layers are composed mostly of quartz, reflecting the fact that it makes up about 98% of the sand on the South Shore barrier beaches. The dark layers have higher concentrations of other minerals such as magnetite (black) and garnet (red).

Hydrology of Barrier Islands

Underneath the surface, barrier islands contain groundwater that fills the pore spaces between the grains of sand. This accumulation, or lens, of freshwater is maintained by precipitation. The water table, which is the subterranean upper surface of the groundwater, has the form of a low mound, with its highest part approximately midway between the ocean and the lagoon. The elevation of the water table approaches sea level toward the shorelines. The highest part of the mound may have an elevation of a few feet above sea level. Underneath this lens of fresh water, the pore spaces are filled by salt water. According to the Ghyben-Herzberg principle, the vertical extent of the freshwater lens below sea level is forty times its extent above sea level. For example, if the highest part of the water table is two feet above sea level, the lens extends to eighty feet below sea level.

At some locations on Fire Island, wells have been sunk deep enough to penetrate below the freshwater lens. But instead of bringing up salt water, these wells flow freely with fresh water without pumping, although the water has a strong taste of dissolved iron and other materials. Each of these wells draws from a confined aquifer, which is under pressure due to one or more layers of clay above it. The source of the water is Long Island itself, and the pressure is provided by its higher elevation than that of Fire Island.

In some swale areas on Fire Island, the surface lies below the water table. Here, wetlands such as marshes, bogs and swamps occur. Some good examples of these wetlands occur at Sunken Forest and Watch Hill.

Cobble Beaches

North Shore beaches and those adjacent to headlands, even along the South Shore, contain a poorly sorted mixture of particle sizes that includes sand and cobbles, rocks with a somewhat flattened shape that prevents them from rolling in the waves. Cobbles are found in smaller numbers on the sandy barrier beaches. These rocks become flatter as they are abraded by sand and gravel due to wave action.

Represented among the cobbles are the same rock types that are included in Long Island's glacial till. Most of these rocks are pieces of Connecticut's bedrock. Quartzite cobbles, composed almost completely of the mineral quartz, are quite common. These are usually translucent and white, grayish or pinkish in color. Rocks that appear completely black are likely to be diabase. However, a look with a hand lens reveals that they are speckled because they contain tiny white grains of plagioclase and dark grains of pyroxene. Banded rocks are gneiss, and the ones that are foliated, or finely layered, are schist. Gneiss usually contains quartz, along with white or pink feldspar and black or silvery flakes of mica. Red garnet may also be present in these rocks. Sedimentary rocks, such as sandstone and conglomerate (which contains pebbles), also occur as cobbles. These two rocks usually contain hematite, which acts as a cement to hold the sand and pebbles together.

Estuaries

Estuaries are places where substantial amounts of fresh and salt water mix. This occurs in bodies of water that are connected to the sea, are partly enclosed by land, and have a source of fresh water such as streams or rivers. Long Island is adjacent to four estuaries: Long Island Sound, Great South Bay, Peconic Bay and the Hudson River, where it enters New York Harbor. Each of these estuaries has its own geologic history.

The lower Hudson River is an example of a fiord, since it is a narrow body of water bound largely by bedrock that was deepened by the scouring action of ice during the most recent episodes of continental glaciation.

Long Island Sound, which is approximately 113 miles long and a maximum of about 20 miles wide, was also scoured out by continental ice sheets, but it is only bordered on the north by bedrock. In the eastern part of the Sound, the maximum depth of water is about 300 feet. The bottom of Long Island Sound is composed of marine sediments underlain by a considerably thicker deposit of glacial material. Much of this substance is clay that was deposited in a freshwater lake that occupied Long Island Sound and Block Island Sound after the ice began to retreat from its last maximum, which occurred about 21,000 years ago. Initially, a line of morainal hills extended east from what is now Montauk Point and created a barrier between the Block Island Sound basin and the Atlantic Ocean. Meltwaters from the glacier were trapped between the ice to the north and the moraine to the south, creating the lake. At this time, another line of hills, extending east from Orient Point, was submerged under the lake waters. An outlet from the lake to the Atlantic existed between the present positions of Block Island and Montauk. As the outlet channel eroded, the lake level lowered, eventually exposing the Harbor Hill Moraine. This separated the lake in Long Island Sound from the waters in Block Island Sound, and a new outlet formed in The Race, just west of what is now Fishers Island. Lewis and Stone (1991) refer to the lake in the Long Island Sound basin as glacial Lake Connecticut. Initially, the surface of this lake was about 33 feet lower than present sea level, however sea level at that time was hundreds of feet lower than today. Gradually, the outlet eroded, ultimately lowering the surface of the lake down to about 200 feet below present sea level. Eventually, the lake drained, exposing the basin as a river valley, with the river fed by streams from Connecticut and what is now Long Island. As the glacial ice was melting, sea level was rising. Around 14,000 years ago, marine waters finally began to enter Long Island Sound.

Along its southern edge, Long Island Sound is, for the most part, adjacent to the glacial moraine of the North Shore of Long Island. The sediment removed from the Sound by the ice sheet probably was mostly Cretaceous sand, clay and gravel, which was subsequently deposited on the continental shelf of the Atlantic Ocean as till and outwash. The portion that is now above sea level is Long Island, but large amounts of the outwash are submerged beneath the Atlantic south of Long Island. The floor of Peconic Bay is composed mostly of glacial outwash of the Terryville Outwash Plain, the surface sediments of which were deposited at about the time when the moraine along Long Island's North Shore was formed.

The sources of freshwater in the estuaries around Long Island are surface water bodies such as streams and rivers, direct runoff from the land, subsurface flow of groundwater, and precipitation that falls directly on the surface of the estuary. Of these, streams, rivers and subsurface flow are the most important.

The volume of seawater that flows in and out with the tides is called the tidal prism. Estuaries differ greatly in the ratio of freshwater input to the tidal prism. The Hudson River supplies a large amount of freshwater to its estuary; therefore the ratio of fresh to saltwater is quite high. The mixing of the fresh and saltwater is incomplete due to the large influx of freshwater. The result is that this estuary is stratified, the less dense fresher water on top and the denser saltwater below. The water does gradually mix, but, as this occurs, additional freshwater is entering, which maintains the stratification. As is true of all estuaries, the Hudson River Estuary is tidal. When the tide rises, the saltwater and its overlying freshwater flow upriver. The falling tide moves this water downstream. Although the direction of water movement oscillates with the tides, the net flow of water is toward the sea.

In Long Island Sound and Peconic Bay, the ratio of freshwater input to the tidal prism volume is much lower, therefore mixing is more complete. However, as is true in all estuaries, salinity varies with location. In the mouths of these estuaries, salinity is nearly 35 parts per thousand, as in the adjacent Atlantic. Where the Peconic River enters Peconic Bay, the salinity is nearly zero. Technically, the estuary only includes the areas where the salinity is between 5 and 35 parts per thousand, but these boundary conditions are somewhat arbitrary. It is best to think of estuaries and other shoreline environments as being part of an integrated whole that includes the environments that are adjacent to them. In fact, even these boundary conditions vary in location with time. They move with the tides and with changing weather conditions that affect rates of stream flow.

The tidal range also varies between and within estuaries. On the Hudson River, the tidal range decreases upstream as the effect of the sea diminishes with distance. In Peconic Bay, the tidal range does not vary much with location. In Long Island Sound, however, the range is about two feet near its eastern end, where it is open to the Atlantic, and increases gradually to the west, reaching over six feet at its western end. This is due to a funneling effect that is caused by its narrowing toward the west.

Salt Marshes

Salt marshes are densely vegetated parts of the intertidal zone. They form in areas that are protected from strong wave action. They often occur in embayments bordered by morainal headlands or where spits of barrier beaches create sheltered lagoons or partially block the entrances to harbors. The largest plant in Long Island salt marshes is Saltmarsh

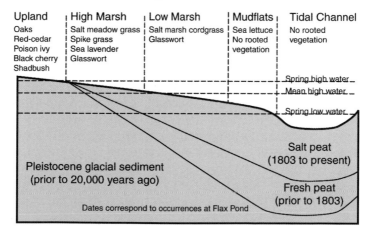

Idealized Cross Section of a Long Island Salt Marsh

Upland	High Marsh	Low Marsh	Mudflats	Tidal Channel
Oaks	Salt meadow grass	Salt marsh cordgrass	Sea lettuce	No rooted
Red-cedar	Spike grass	Glasswort	No rooted	vegetation
Poison ivy	Sea lavender		vegetation	
Black cherry	Glasswort			
Shadbush				

Spring high water
Mean high water
Spring low water

Salt peat
(1803 to present)

Pleistocene glacial sediment
(prior to 20,000 years ago)

Fresh peat
(prior to 1803)

Dates correspond to occurrences at Flax Pond

Cordgrass (*Spartina alterniflora*). It can grow over five feet in height, but it is often shorter. It spreads to new areas by means of seeds. Once it is established in an area, its predominant mode of reproduction is rhizome spreading. This grass has physiological mechanisms which allow it to rid its tissues of excess salt. This enables it to grow in areas that are covered by salt water during high tide.

The vegetation in a Long Island salt marsh occurs in zones that are determined by the tides. The lowest parts of the salt marsh are always covered by water and are called tidal channels. These are bordered by mudflats which are mostly devoid of rooted vegetation and are exposed to air only during the lowest tides. Masses of rockweed and other algae often grow here. The otherwise barren mudflats are interrupted by tussocks, clumps of *Spartina alterniflora* and Ribbed Mussels. These tussocks may have originated as pieces of the low marsh that were carried out onto the mudflats by ice that was rafted about by the tides. Like storms, episodic occurrences of thick ice may significantly alter the surface of the marsh by redistributing sediment.

Adjacent to the mudflats and higher in elevation is the zone where Saltmarsh Cordgrass grows. This is called the low marsh and its upper limit, as far as elevation is concerned, corresponds to the average height of the high tide (mean high water). The tallest *Spartina alterniflora* grows where the low marsh meets the mudflats.

Above the low marsh is the high marsh, which is bounded by the mean high water and spring high water lines. This area is characterized by a mixture of Saltmeadow Grass (*Spartina patens*) and Spike-grass (*Distichlis spicata*) which are generally shorter and not as upright in appearance as the *Spartina alterniflora* which grows in the low marsh. At first glance, the high marsh appears to be a soft, comfortable place to lie down. However, it is always soggy and is actually covered by water during higher than average high tides.

Often, portions of marshes are covered by straw-colored dead stems of Saltmarsh Cordgrass. Bare patches are often created by these rafts, which smother any vegetation upon which they come to rest. This type of bare area is frequently filled with holes made by fiddler crabs. These areas tend to be colonized first by Slender Glasswort (*Salicornia europaea*).

The stands of grass that grow in the salt marsh slow down the movement of the seawater as it is brought in by rising tides. This causes suspended silt to trickle out of the water and become deposited on the marsh as mud. Gradually, the roots and rhizomes of the grasses bind the sediment together to form a spongy substance called peat, which makes up a large part of the marsh surface. The vertical accumulation rate of peat is variable, but averages about three millimeters per year. This is about equal to the current rate of sea level rise.

At Flax Pond and some other salt marshes, the saltwater peat is underlain by a layer of older peat that was deposited in freshwater. This indicates that these marshes were formerly beyond the reach of the tides, enabling freshwater wetlands to exist there. In the case of Flax Pond, the inlet connecting it to Long Island Sound did not exist before 1803. In that year, it appears that residents in the Old Field area dug the inlet in order to transform Flax Pond into a site where oysters and clams could be harvested.

Other marshes, such as Bass Creek on Shelter Island, contain buried freshwater peat that reflects a time when sea level was lower than it is today. In fact, at Bass Creek, the freshwater peat overlies an ancient soil horizon that probably developed in a former woodland that existed at the site. When considered together, this evidence tells the story of a rising sea level, which gradually flooded a forest by causing the water table to rise. This resulted in a freshwater wetland. Eventually, this wetland was invaded by tidal waters, first on an occasional basis, and later at regular intervals. The site then became the salt marsh that it is today.

At Hubbard County Park on Peconic Bay, a salt marsh lies behind a protective beach that formed as a spit. On the bay side of the beach, freshwater peat can be found in the intertidal zone. This peat indicates that when sea level was lower, the spit was located seaward of its present location. As sea level rose, the beach gradually migrated landward along with the shoreline and eventually overrode the marsh. As it continued to migrate, the peat became exposed on the beach face.

CHAPTER II

Seashore Ecology

Long Island's saltwater communities host a wide array of different plants and animals that are adapted to particular habitats. The distribution of each species varies, some being more tolerant to varying environmental factors and others requiring very specific niches. The external conditions which influence where marine life can exist include temperature, salinity, substrate, currents, water depth, tides and light. This chapter provides an in-depth look at the common biota that reside in each habitat as well as their interaction with the environment and other species.

The most rewarding time for wildlife observation at any marine location is at low tide, when the intertidal zone is exposed. Appropriate attire should be worn, especially boots, as the areas covered by water at high tide, such as the lower parts of the salt marsh, will be wet and usually muddy. Also, for bird observation, binoculars are strongly recommended.

Virginian Faunal Subprovince

The coastal waters of the world are divided into a human-made classification of regions, and, in turn, into smaller provinces, mainly influenced by water temperatures. Long Island lies within the American Atlantic Temperate Region which extends from Cape Cod, Massachusetts to Texas. The region from Cape Cod to Cape Hatteras, North Carolina is more specifically placed in the Virginian Faunal Subprovince. Southern and northern marine species will migrate into our subprovince but do not become residents as they are unable to endure the extremes of winter and summer, respectively.

Food Web

In nature, one fundamental requirement for living things is their need for nourishment. A few plants and plantlike animals manufacture their own sustenance and are known as autotrophs, but the vast majority of life forms are heterotrophs and require food from external sources. In order to guarantee a species' survival, the habitat where it lives must provide this source of nutrition. The interaction and dependence of all life on each other for nourishment within various ecosystems forms an intricate matrix called the food web.

The Food Web

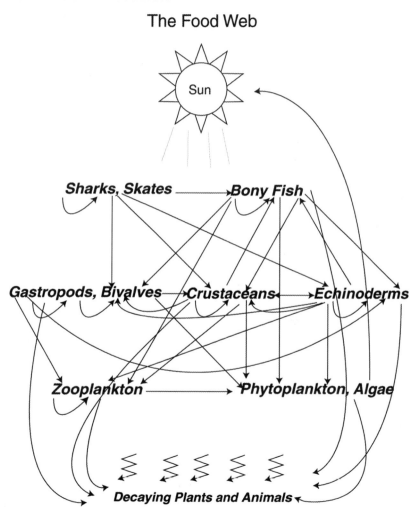

The food web is comprised of organisms that are classified into trophic levels based solely on their feeding behavior. The lowest and first level contains the producers, such as plankton and green plants which, fueled by the sun's energy, convert nutrients into living material through photosynthesis. This is followed by the herbivores which are the plant eaters. The third level consists of the primary carnivores, meat eaters which feed on the herbivores, while the fourth level is reserved for the secondary carnivores which eat the primary carnivores. An organism's position within the food web is often not well defined, as in the example of the omnivores which feed on plants and animals and can be classified within several trophic levels. The decomposers, a separate trophic level, are categorized by organisms which break down decaying matter into nutrients often used by the producers.

Plankton and Nekton

Plankton consists of aquatic plants and animals which are usually unable to make strong independent movements and are transported by tide and current. Essential components of the food web, plant plankters are called phytoplankton and animal plankters are called zooplankton. The composition of the plankton varies seasonally, usually including a spring bloom for phytoplankton, an increase of zooplankton in the summer, and a second bloom in the fall.

Classification of plankton is also governed by body size and is broken down accordingly:
Ultraplankton - 5 microns
Nanoplankton - 5-60 microns (ultraplankton and nanoplankton include organisms such as diatoms, dinoflagellates, algae, fungi and bacteria)
Microplankton - 60-1000 microns
Mesoplankton - 1-5 mm (microplankton and mesoplankton include protozoans, herbivorous larvae, juvenile crustaceans and appendicularians)
Macroplankton - 5 mm (fish larvae, small crustaceans and comb jellies)
Megaplankton - larger than 5mm (jellyfish, comb jellies)

Phytoplankton includes such one-celled plants as diatoms, dinoflagellates and some blue-green algae. Diatoms are plant organisms that have silicate skeletons. Unlike diatoms, dinoflagellates have cellulose cell walls and also have a whiplike "tail" for swimming. The well known "red tide", pinkish dinoflagellates, is toxic to fish and other animals.

Long Island's "brown tide" is caused by the phytoplankton *Aureococcus anophagefferens* and blooms in the nutrient-rich waters of our inland bays. This phytoplankton kills off local shellfish, such as the Atlantic Bay Scallop (*Argopecten irradians*), by blocking their filtration systems and also by crowding out other phytoplankton used for food by the shellfish.

Zooplankton, the minute drifting animals of the plankton, are further subdivided into the holoplankton, permanent members of the community, and the meroplankton, animals that spend only a portion of their life cycle as zooplankton. The majority of the holoplankton are represented by the copepods, marine crustaceans whose species are mainly herbivorous, feeding on phytoplankton. *Calanus finmarchicus*, the most abundant of the copepods, is an important food for herring and mackerel. Other holoplankton include Chaetognatha (arrow worms), appendicularians, horned krill shrimp, pteropods and numerous others. The pteropods, also known as sea butterflies, are one of the few mollusks that are found in the holoplankton, and whose tiny, delicate, translucent shells infrequently wash ashore.

The meroplankton consists of the larval stages of coelenterates, mollusks, crustaceans, and fish. These members of the zooplankton frequently do not have any resemblance to the adult species. An advantage of a different, mobile body form among the plankton is easier species dispersal. One such example is the Eastern Oyster (*Crassostrea virginica*), which has two stages of development in the zooplankton; it hatches as a free-swimming, ciliated trochophore, later becoming a more developed veliger.

Nekton are the animals who are strong, free swimmers, and are not dependent on the current. Squids, cartilaginous fish (sharks and their allies), bony fish, horseshoe crabs, sea turtles and whales are included in the nekton.

Salt Marsh Community

The salt marsh is a habitat with plants and animals uniquely adapted to its conditions and is a haven and source of nutrition for many other marine animals, such as young fish and invertebrates. The plants that inhabit the salt marsh are tolerant of higher salinity due to the physiological adaptations they have evolved, enabling the animals to gain nourishment from the increased levels of salt in the edible plants. As the plants die, organic materials produced from the decayed vegetation deposit into layers of peat. During high tide, the water draws off some of the organic matter and bacteria further breaks down the plant fragments into chemical compounds. These are distributed to the coastal water and utilized by marine plants, especially phytoplankton.

The habitat of a fully developed pocket salt marsh, such as Flax Pond, is subdivided into the low marsh, high marsh and the scrub-shrub wetland, a transition zone between the high marsh and the upland. There are also fringing salt marshes along Long Island's shores. An example is at the water's edge of Long Beach Bay at Orient Beach State Park, where the flora and fauna found at this location are typical low marsh species.

Plants of the Salt Marsh

At the salt marsh's edge, sediments deposit, creating a habitat ideal for one of the primary and indicative inhabitants of the low marsh, *Spartina alterniflora*, Saltmarsh Cordgrass. *Spartina's* cell structure is adapted for living in the saltwater that floods the marsh twice a day. The tall, erect grass aids in additional deposition by anchoring the sediment within the area where it grows. Interspersed among the Saltmarsh Cordgrass is Sea Lavender (*Limonium carolinianum*), which produces tiny, purple flowers, and Glassworts (*Salicornia* spp.), with fleshy, jointed, green, erect stems which turn red in fall. *Salicornia*, from the Greek meaning "salt horn", is an edible plant that has a pleasant salty taste and can be used to make salt pickles. It is an important food source for geese and ducks. Salicornia is a member of the goosefoot family which also includes spinach and beets.

The high marsh, landward of the low marsh, is flooded irregularly by high tides. Another cordgrass dominates this region, *Spartina patens*, Saltmeadow Cordgrass or salt hay, a matted, flattened, grass. Also present is Spike-grass (*Distichlis spicata*), very tolerant to high salinities, and Black Grass (*Juncus gerardii*), a rush, commonly found bordering the high marsh and the scrub-shrub wetland.

Bordering the salt marsh is an invasive plant called the Common Reed (*Phragmites australis*). Though common in other habitats where conditions have been disturbed, i.e. by fertilizers, this species is particularly obtrusive in the salt marsh. Here it outcompetes some of the native plants which are more ecologically valuable to the ecosystem.

The scrub-shrub wetland, a transition zone between the high marsh and the upland, is dominated by Marsh Elder (*Iva frutescens*) and Groundsel Bush (*Baccharis halimifolia*), whose fall blooms are white, cottony tufts. Also present are Poison Ivy, Beach Rose and Seaside Goldenrod.

Animals of the Salt Marsh

Numerous animals, including herons and egrets, ducks, muskrats and raccoons, fish, rails, gulls, and other seabirds, visit the salt marsh for their source of food. The following paragraphs discuss some of the important and indicative residents of this community.

Found at the intertidal zone of the salt marsh is the Ribbed Mussel (*Geukensia demissa*), an abundant bivalve with an extremely high tolerance to a wide range of salinity. At low tide, colonies of Ribbed Mussel protrude from between the Saltmarsh Cordgrass along the low marsh, their byssal threads entwined with the *Spartina* roots.

The Fiddler Crab (*Uca* spp.), a small crustacean, is named for the male's enlarged claw, which is used for fighting and is waved about to attract females. There are two species found on Long Island and their habitats overlap within the salt marsh. These crabs are burrowers and numerous burrows are common throughout the intertidal zone. The fiddler crab will scurry in and out of their burrow during low tide. They are a source of food for the Diamondback Terrapin, herons, egrets, rails and raccoons.

If you look closely at the grasses in a salt marsh, you may see the small, inconspicuous Eastern Melampus (*Melampus bidentatus*), a pulmonate snail that has a lung and breathes air. Found at the base of the grasses at low tide, the snail has an internal clock which enables it to anticipate high tide and climb the stalks ahead of the rising water. *Melampus* is capable of holding its breath long enough to survive while the grass is submerged. Many birds and invertebrates rely on this snail as a source of food, including the Marsh Wren, muskrats, egrets, plovers and yellowlegs.

Other residents include the Mummichog (*Fundulus heteroclitus*), a snail-eating fish which remains in tidal pools during low tide, the Salt Marsh Mosquito (*Aedes sollicitans*), whose adult females are blood-suckers and whose larvae, called wrigglers, live under the salt marsh surface, and the carnivorous Diamondback Terrapin (*Malaclemys terrapin*), the only turtle who resides exclusively in the salt marsh.

FreshWater Tidal Marsh Communities

Long Island also has freshwater tidal marsh communities, where the waters ebb and flow in a manner similar to the waters of the salt marsh. An example of an area on our island where water fluctuation of this type occurs is at the Nissequogue River, and its inhabitants are typical freshwater wetlands species. This book will not list the biota found at this location, but a good reference for the species found in this habitat is *A Field Guide to Long Island's Freshwater Wetlands*.

Estuarine Communities

Seawater mixes with freshwater, producing specialized habitats called the estuarine community. The concentration of salt in this mixture, the salinity, determines which plants and animals reside in a particular type of estuary. Also a factor in the biota of these communities is the substrate, as certain species need to attach to a hard surface, while others may require a soft substrate for burrowing. Lastly, the depth of the body of water influences which species of plants and animals are able to reside there.

Estuarine communities are well represented along Long Island's coast by Long Island Sound, Great South Bay (a lagoon), and Peconic Bay. There are different niches within each of these bodies of water, and the species vary within these locations.

Starting at the beach and proceeding into the estuarine waters, the first habitat encountered is the intertidal zone. The marine life of this region is ecologically subdivided into: the benthic epifauna, surface animals attached to the substrate; the mobile epibenthos, bottom dwellers freely moving on the surface; the benthic infauna, animals that live under the sand and mud; and the intertidal shorebirds.

The next area, the subtidal zone, consists of the deeper water outside the intertidal zone, represented by the benthic species, the megazooplankton, the nekton, and the diving and aerial birds. The eelgrass community is also present in the subtidal zone, but this environment is so unique, encompasses so many different species, and is found in both the estuaries and ocean, that it will be discussed in a separate section of this chapter.

Plants of the Estuarine Communities

The intertidal flats support a large quantity of one-celled plant life, including microalgae, diatoms, and blue-green algae, which can form dense mats at the high intertidal zone. Depending upon the species' characteristics, the benthic microfauna either remain adhered to the surface, or move up and down in the sediment.

Macroalgae also are represented in the intertidal and subtidal zones by the seaweeds. Individuals attach to hard substrate such as rocks, shells, jetties, seawalls, and bulkheads. Some common species are Sea Lettuce (*Ulva latuca*), often found floating or washed ashore, Bladder Rockweed (*Fucus vesiculosus*), and Green Fleece (*Codium fragile*). This plant life extends into the subtidal zone, but the depth of water it inhabits is limited by the plants' need for sunlight.

Animals of the Estuarine Intertidal Flats

Benthic Epifauna
The rocky shores, or other locales with hard substrate, provide a suitable habitat for the animals that remain attached to the surface. These species, once fixed in their position, usually remain there throughout their lives and are food for numerous animal species. Typically, they are filter feeders as they are unable to forage for food.

The most notable example of benthic epifauna is the Eastern Oyster (*Crassostrea virginica*), an estuarine bivalve. *Crassostrea* thrives in brackish water, as it is very tolerant of fluctuations in salinity. Beginning life as a mobile component of the zooplankton, the oyster larva eventually cements itself to a hard surface, often an old oyster shell, where it remains through its maturity. An oyster's sedentary nature makes its survival rate low, as it is easy prey to oyster drills, sea stars, flatworms, rays and oystercatchers. Its shell can be damaged by boring clams, sponges, and mud worms, and the animal can be parasitized by fungus. Residing in areas of lower salinity eliminates the chance of predation, but alas, the bivalve never reaches it full size or productivity in such an environment. Our local species is commercially harvested, and as water quality in locations such as Hempstead Harbor improves, the oyster is making a resurgence in Long Island's waters.

The abundant Blue Mussel (*Mytilus edulis*) is another bivalve which attaches to hard surfaces such as pilings, jetties, rocks, shelly bottoms and piers. *Mytilus* attaches itself by tough byssal threads and, unlike the oyster, can detach itself and move to a different location. This mussel can be found in a wide range of locations in the estuarine community, from slightly brackish to deep offshore waters, as it has a high tolerance to different salinities. It is a favorite food for invertebrates and vertebrates, including humans, and is commercially harvested.

Other species of this niche include the Ivory Barnacle (*Balanus eburneus*), with its calcareous, volcanolike exoskeleton cemented to the substrate, the Common Atlantic Slipper Shell (*Crepidula fornicata*), with individuals that live stacked upon one another, and the Jingle Shell (*Anomia simplex*), a bivalve that attaches to the hard surface with a stemlike peduncle.

Mobile Epibenthos
On the mud and sand flats of the intertidal zone, there are bottom dwellers that move about freely on the surface. Their mobility enables these species to forage for food, whether they are predators or scavengers.

The Eastern Mudsnail (*Ilyanassa obsoleta*), a small, abundant gastropod usually covered with mud, algae, and debris, is found on flats and intertidal creeks in large quantities. A scavenger that also feeds on bottom sediments rich in tiny algae, it has the special ability to detect dead organisms, with hundreds of individuals descending upon the carrion. During the winter months, the snails move into the deeper waters of the estuaries. They are a food source for some fish, waterfowl, shorebirds and muskrats.

Another common scavenger that feeds on algae is the Long Clawed Hermit Crab (*Pagurus longicarpus*). Unlike other crabs, *Pagurus* has a soft, elongate body which it protects by inhabiting empty snail shells. Hermit crabs must relocate to a larger shell as they grow, and are often combative with each other as they compete for those shells. Besides occasionally eating each other during combat, the crabs are food for many fish and wading birds. A unique commensal arrangement occurs between *Pagurus* spp. and the hydroid *Hydractinia echinata*, commonly known as Snail Fur. *Hydractinia* covers the crab's shell, its stinging cells deterring predators from attacking the hermit crab, while the movement provided by the crab aids the hydroid in getting its food and water.

The Common Spider Crab (*Libinia emarginata*) is a slow-moving crustacean found in shallow estuaries or lagoons on muddy, sandy or shelly bottoms. For camouflage *Libinia* covers itself with living algae plants, debris, and small animals such as hydroids, bryozoans, and sponges, by attaching them to hairs on its carapace. The spider crab is believed to be a scavenger and is prey for other bottom dwellers, such as cod.

Mobility in the intertidal zone is essential for predation and Forbes' Sea Star (*Asterias forbesi*) is a carnivorous echinoderm found in this niche. This starfish tolerates the lower salinities found in estuaries and moves along the bottom on its tube feet, or floats freely with the current. *Asterias* also uses its tube feet to pry open bivalves by wrapping its arms around both valves, steadily exerting pressure. Once the shells are slightly ajar, the starfish inserts its stomach into the opening where it digests the mollusk's soft tissue. Forbes' Sea Star feeds on oysters, clams, mussels, snails, and barnacles. Adult sea stars do not seem to have any predators but they are prey as larvae.

Estuarine molluscan predators include the Atlantic Oyster Drill (*Urosalpinx cinerea*), Thick-lipped Oyster Drill (*Eupleura caudata*), Knobbed Whelk (*Busycon carica*), Channeled Whelk (*Busycotypus canaliculatus*) and Shark Eye Moonsnail (*Neverita duplicata*). The drills and the moonsnail use their radulae to drill a hole through the shell of their prey. A whelk, like a sea star, will pry the shells of its victim open by exerting pressure with the strength of its foot wrapped around the valves.

Benthic Infauna

Infaunal animals live under the mud and sand flats in the intertidal zone of the estuarine community. The composition of the sediment where they reside must be soft for them to dig their burrows, and gently moving seawater must cover their location for them to feed. Mollusks, sandworms, and mantis shrimp are common infaunal species.

The bivalves residing under the sediment all possess a fleshy foot for digging, in addition to a pair of siphons, one inhalant, the other exhalant, that take in water and food and expel water with waste, respectively. At high tide, these mollusks extend their siphons out to feed, the length of their siphons usually determining the depth of their burrows. All Long Island infaunal bivalves are filter feeders, except for the Baltic Macoma (*Macoma balthica*), which is a deposit feeder. Examples of burrowing clams include the Common Razor Clam (*Ensis directus*), the Soft-shelled Clam (*Mya arenaria*) and the Northern Quahog (*Mercenaria mercenaria*). These are valued food for raccoons, shorebirds, waterfowl, crabs and predacious gastropods.

A voracious predator, the Clam Worm (*Nereis virens*) lives as benthic infauna in a mucus-lined burrow. Found in coastal bays and brackish estuaries, it hunts on the surface of the intertidal flat during high tide, feeding on other worms, crustaceans, small fish, carrion, and algae. Fish, rays, and shorebirds feed on *Nereis*, and it is popularly used by fishermen as bait.

The elusive Common Mantis Shrimp (*Squilla empusa*) is a fierce, carnivorous predator, with a claw similar to that of a praying mantis. Located in the lower intertidal zone on mud bottoms of estuaries, this nocturnal shrimp hides in a burrow with multiple exits. Its diet consists of crustaceans and other invertebrates.

Shorebirds

Certain species of birds feed at the shoreline of the estuarine intertidal zone, preferably at low tide, when their food source is exposed. They have adopted specialized feeding methods to obtain their nourishment. The Ruddy Turnstone (*Arenaria interpes*) turns over rocks with its short, stout beak, hunting for Beach Fleas and other small invertebrates. A shorebird that probes the sand with its bill for prey, the Black-bellied Plover (*Pluvialis squatorola*) is known to run in short bursts. The Lesser Yellowlegs (*Tringa flavipes*) runs about the mud picking up exposed snails, crustaceans and worms. A most unique forager is the exclusively intertidal American Oystercatcher (*Haematopus palliatus*), which breaks open bivalves with its beak.

Estuarine Subtidal Animals

Nearly all of the marine animals represented in the intertidal estuarine community will also inhabit various depths of the subtidal zone. The extent of their range into the deeper waters is different for each organism, and those depths are provided with the species' descriptions.

Benthic Subtidal

In addition to the previously mentioned benthic intertidal species, there are animals that primarily spend their adult existence in the benthic subtidal zone. One such arthropod is the Northern Lobster (*Homarus americanus*), and its subtidal range can be as near as the shoreline to as deep as the continental shelf. An omnivore, feeding on prey, carrion, or plant material, it is located in the sound and bays, as well as oceans. There has been a recent die off of lobsters in Long Island Sound, and the cause of their rapid decline is unknown at the present time. Scientists are currently studying the Sound's lobsters to solve this problem.

The Horseshoe Crab (*Limulus polyphemus*), a primitive arthropod, also resides as an adult in subtidal waters. Mature, large horseshoe crabs aggregate in the deeper estuarine waters most of the year, while the smaller juveniles inhabit the shallower waters. The Horseshoe Crab is a predator, plowing through the sediment for worms and mollusks. In May and June, during high spring tides, *Limulus* will come up to the shore, in great numbers, to lay their eggs in the upper limits of the sandy intertidal flats exposed at low tide. During the evening hours on the beaches of Long Island Sound, it is quite common to see many Horseshoe Crabs amassing near the shoreline. Their eggs are vulnerable to predation and fall prey to numerous animals such as gulls and terns.

Megazooplankton

Borne by the tide and the current, floating in the estuarine subtidal waters, the megazooplankton are represented by the jellyfish.

Leidy's Comb Jelly (*Mnemiopsis leidyi*), a ctenophore, tolerates low salinities within the estuarine community. Easily recognized by the internal combplates that produce prismatic colors inside their transparent bodies, they will flash green when they are disturbed. *Mnemiopsis* does not sting and has short tentacles, feeding mostly on small organisms and larvae.

The two coelenterates commonly found in these waters are the Moon Jellyfish (*Aurelia aurita*) and the Lion's Mane Jellyfish (*Cyanea capillata*). *Aurelia* is tolerant of lower salinities, produces venom that is harmless to humans, and feeds on small planktonic animals. The Lion's Mane Jellyfish swarms in Long Island's bays and estuaries from the spring through the summer, has stinging tentacles that produce a nasty, itchy rash, and feeds on fish.

Nekton

The fish are the largest group of free-swimming animals within the estuarine communities. All the species discussed below are popular sports fish, except the silverside, which is valuable as bait.

The most abundant of all of the nekton found in estuarine communities is the Atlantic Silverside (*Menidia menidia*). The slender, streamlined *Menidia* lives in shallow water, and the bays become alive with schools of silversides in June. An omnivore, it eats small fish, invertebrates, detritus and algae. Terns, snappers, striped bass, flounders and other fish feed on *Menidia*.

Two species of flatfish occur in Long Island's waters: Winter Flounder (*Pleuronectes americanus*) and Summer Flounder, or Fluke, (*Paralichthys dentatus*). Both fish are carnivorous and camouflage against the bottom sediment. Winter Flounder, a right-eyed flatfish, inhabits muddy-sand bottoms, occasionally dwelling on sand and clay.

Found in spring and fall in our estuaries, it summers in deeper water. Fluke, a left-eyed flounder larger than the previous species, spawns in ocean waters, but its young move into the shallow, estuarine waters in the summer. It feeds on small fish and shrimp.

Another inhabitant is the Bluefish (*Pomatomus saltatrix*), whose young, called snappers, move into the shallow estuarine waters from late May to November to feed on silversides, killifish, crustaceans, snails, and annelids. Mature Bluefish spawn in the deep ocean waters of the continental shelf.

Striped Bass (*Morone saxatilis*) is a fish that spawns in the Hudson River and then migrates between the estuaries and the river from April to December. Small "Stripers" are found in April and May and larger specimens in late May and June. In fall, schools of immature Striped Bass return to the river for the winter, while larger, mature *Morone* move into deeper ocean waters. A carnivorous predator, it feeds on small fish, crustaceans, annelids and insects.

Aerial and Diving Birds

The subtidal estuarine community includes birds that fly, surface dive, swim, or skim to find their food. Terns dive from the air for their prey, nabbing fish close to the surface. Gulls do not dive for food but usually drop it from above onto hard surfaces to break the shells of their prey.

The Black Skimmer (*Rynchops niger*) flies low over the water while feeding and grabs its food with its unusually long lower mandible. The Double-crested Cormorant (*Phalacrocorax auritus*) sits low in the water, then surface dives for fish. Another fish eater, the Common Loon (*Gavia immer*) swims, catches and swallows its prey underwater.

The majestic bird-of-prey, the Osprey (*Pandion haliaetus*), exclusively a fish eater, is a migratory hawk found around Long Island's estuarine waters from April to October. The Osprey hunts high over the water, spots its prey, dives talons first and grabs the fish in its claws. Once near extinction in our area due to the insecticide DDT, *Pandion* is becoming more common and can frequently be observed on high, human-made, nesting platforms.

Eelgrass Communities

In beds that range in size from a few sparsely clumped plants to dense underwater "forests", Eelgrass (*Zostera marina*), a flowering plant, lives totally submerged in its seawater environment. Eelgrass is an important contributor to its surroundings as its root system provides sediment stability beneficial to benthic animals, and its long, ribbonlike leaves offer a primary habitat for many local organisms, in addition to shelter for adult and immature invertebrates and fish.

Zostera grows mostly in the shallow sublittoral zone on both muddy and sandy bottoms in Long Island Sound, Great South Bay, Peconic Bay, and lagoons behind other barrier beaches. Its tolerance to salinity allows Eelgrass to inhabit both estuarine and marine waters, providing that the water currents are not too turbulent.

Plants of the Eelgrass Community

Other than the *Zostera marina* that dominates this community, the plantlife found in this habitat is non-flowering and epiphytic, either growing exclusively on Eelgrass, or attached to inorganic materials such as rocks and shells. The phytoplanktonic diatoms can coat the *Zostera* leaves, but the most dominant plants are the algae, especially the seaweeds. These can include species of stoneworts and blue-green, red, green and brown algae.

Animals of the Eelgrass Community

Epiphytes also include any sessile organisms that live attached to Eelgrass leaves. Spiral Tube worms (*Spirorbis* spp.), Spiral-tufted Bryozoan (*Bugula turrita*), barnacles and encrusting sponges are a few examples of these types of animals.

Grazers and foragers are the mobile animals which crawl along the surface of *Zostera*, exemplified by the herbivorous gastropods. The predominant species is the abundant Alternate Bittium (*Diastoma alternatum*), a tiny snail, with a shell only 5 mm. long, that feeds on algae and diatoms along Eelgrass leaves.

Benthic epifauna live among the base of Eelgrass on muddy or sandy bottoms. The Atlantic Bay Scallop (*Argopecten irradians*), a filter feeding bivalve, primarily lives attached by its byssal threads to the *Zostera*. While the scallop feeds, its valves are agape, exposing a mantle edged with dozens of light-sensitive, blue, iridescent eyes. When disturbed, *Argopecten* becomes mobile by removing its byssus and quickly opening and closing its shells, swimming in erratic, jerky movements through the water. In recent years, the bay scallop has not been plentiful due to the brown tide, a massive bloom of a particular species of phytoplankton. Currently, studies are being conducted to determine the cause of this phenomenon.

The Blue Crab (*Callinectes sapidus*), an aggressive, predacious crustacean lurking in the cover of Eelgrass beds, is a member of the mobile epibenthos. *Callinectes* bottom-feeds on worms and invertebrates but will eat almost anything dead or alive. The animal itself provides a habitat for barnacles and nemertean worms. When young, blue crabs are especially vulnerable to predation by Sand Sharks, fish, gulls and wading birds.

Two closely related bony fish, the Northern Pipefish (*Syngnathus fuscus*) and the Lined Seahorse (*Hippocampus erectus*) are adapted for living within *Zostera* beds. Pipefish hide among plants, often vertically aligning their thin, elongate bodies with Eelgrass leaves, while seahorses use their coiled, prehensile tails to wrap around the blades. Both have a long, tubular, snout with a toothless mouth and feed on shrimp, other small crustaceans and fish larvae. Males have a brood pouch, a very unique feature, where the females' eggs are deposited. They are prey to Blue Crabs and larger fish.

Sticklebacks are another type of bony fish found in this environment, especially common in the Great South Bay. The Threespine Stickleback (*Gasterosteus aculeatus*), larger with narrow plates on its body, and the Fourspine Stickleback (*Apeltes quadracus*), smaller with a scaleless body, spawn in freshwater streams and are famous for their unusual nesting habits and the care of their young. Their diet consists of small crustaceans, while they are eaten by larger fish and wading birds.

Shallow water cartilaginous fish, including the Sand Shark (*Carcharhinus plumbeus*) and the Cownose Ray (*Rhinoptera bonasus*) feed in Eelgrass beds. The Cownose Ray has destructive foraging habits, uprooting and damaging *Zostera* plants while hunting for clams. The Sand Shark is a predator and a scavenger, preferring to eat young Blue Crabs.

Atlantic Ocean Ecology

On Long Island's oceanic coast, there are two types of intertidal environments: the sandy shoreline of the barrier beaches, and the rocky coastline found on the eastern tip of the South Fork. Each supports a separate, distinctive ecosystem with a characteristic marine life. Locally, beaches at Jones Beach State Park and Fire Island National Seashore exemplify Long Island's sandy outer beaches, while the shores of Montauk Point State Park represent Long Island's rocky oceanic shoreline. The sandy shores constitute the overwhelming majority of Long Island's Atlantic Ocean coastline.

The deeper water subtidal region is home to its own array of unique marine animals that are represented by the benthic epifauna and infauna, mobile epibenthos, nekton and aerial birds.

Ocean Plants

With the exception of the phytoplankton floating in the ocean waters, plantlife within the sandy intertidal zone is nonexistent. There is no flora that can tolerate the extreme conditions of rapidly shifting sands and harsh waves and currents.

However, the rocky intertidal zone of the ocean waters, similar in its makeup to the coastal regions of New England, does provide sufficient stability to support plants. Seaweeds are the prevalent vegetation in this environment, and the most commonly found is Bladder Rockweed (*Fucus vesiculosus*). *Fucus*, like many other seaweeds, anchors itself to rocks or other hard substrate by means of a holdfast.

Animals of the Sandy Intertidal Zone

Animal life, albeit very few species, inhabit the sandy intertidal zone of the Atlantic Ocean. One species is dominant within this community, the Mole Crab (*Emerita talpoida*). This crustacean, residing in the swash zone, is an amazing example of physiological and behavioral adaptations to an environment. Ovoid in shape and sand colored, its back legs are incredibly well suited for digging, and it uses its tail to anchor itself under the sand.

After a wave breaks, Mole Crabs quickly unearth themselves from under the sand. They then scurry to a new location in the backwash and dig themselves into the sand, their backs toward the ocean. When the next wave crashes and the backwash reappears, the crabs unfurl their feathery antennae to catch plankton or bits of organic matter in the water. Mole Crabs stay in this position until they sense the tide either ebbing or rising, then, en masse, reposition themselves in the sand.

Emerita rapidly burrows in the sediment to avoid predation from shorebirds, fish, and Blue and Lady Crabs. During mating season, the smaller males remain semi-parasitic on the larger females and stay attached for long periods of time. Mole Crabs move to deeper water in winter.

Infrequently occurring from the Rockaways to Fire Island is the Coquina (*Donax variabilis*), a burrowing bivalve. Ocean currents from the south carry this bivalve's larvae to Long Island's shores, but the adult, unable to overwinter in our region, only survives through the warm season. Specimens found here are smaller and darker in shell color than the Coquinas located on the southeastern coast of the U.S., the region where this species is the dominant animal of the sandy intertidal zone. *Donax* is a filter feeder, and it randomly burrows and re-emerges from the sand.

Sanderlings (*Calidris alba*), colonial shorebirds, are commonly found in the swash zone, scurrying up on the beach in front of an advancing wave. They probe the sand with their thin bills for small invertebrates, such as the Mole Crab. Not exclusive to ocean beaches, *Calidris* can also be spotted in the swash zone of estuarine beaches.

Animals of the Rocky Intertidal Zone

Only found on Long Island east of Moriches Inlet, on rock jetties or Montauk's rocky coast, the Atlantic Dogwinkle (*Nucella lapillus*) is at the southernmost extreme of its range. As part of the mobile epibenthos, this carnivorous, predacious gastropod employs its radula to drill a hole in its prey, which primarily consists of barnacles and Blue Mussels.

Unlike the dogwinkle, the range of the Northern Barnacle (*Balanus balanoides*) extends throughout Long Island's ocean shores. Exclusively marine, this barnacle has a narrow tolerance to salinity. As an immobile species of the benthic epifauna, the Northern Barnacle competes with Blue Mussels and rockweeds for space on rocks, jetties, and pilings. *Balanus balanoides* is a filter feeder, and the barnacle's feathery legs extend out between the two calcareous plates atop its conical exoskeleton.

Atlantic Ocean Subtidal Zone Animals

Though located intertidally in the waters of New England, the Northern Moonsnail (*Euspira heros*) resides only subtidally in the ocean off Long Island. Utilizing its huge, fleshy foot that almost completely envelops the gastropod's shell, this animal plows just barely under the surface in search of bivalves. Shells washed ashore are evidence of the snails' presence, as are their sand collars, a circular mix of eggs, sand and dried mucus that disintegrates when dried.

Related to the Blue Crab, the Lady Crab (*Ovalipes ocellatus*) is both a scavenger and a predator. A swimming true crab, *Ovalipes* moves in with the tide to feed, digging itself into the sand with only its eyes visible. Like a corkscrew removing a cork, this crab pushes its well-developed claw into a Mole Crab bed, grabs one, runs in a circular motion holding onto the victim, and gradually pulls the Mole Crab out of its burrow. The Lady Crab is a favorite food of seagulls.

The Atlantic Surf Clam (*Spisula solidissima*), the Arctic Wedge Clam (*Mesodesma arctatum*) and the Smooth Astarte (*Astarte castanea*) are filter-feeding bivalves exclusively inhabiting the oceanic subtidal zone.

Those black, leathery, rectangular egg cases, called mermaid's purses, washed up on shore, are evidence that the Clearnose Skate (*Raja eglanteria*) inhabits the ocean water of Long Island. A cartilaginous fish related to the sharks, the Clearnose Skate has a thick tail with spines on its dorsal surface. *Raja* is a bottom feeder, crushing mollusks and crustaceans with its powerful teeth. Commonly located inshore, this skate relocates to deeper water in the winter.

The free-swimming marine nekton also includes Weakfish (*Cynoscion regalis*), a large mouthed fish, and Bluefish (*Pomatomus saltatrix*), a strong jawed, sharp toothed fish; both reside exclusively in marine and estuarine saltwater. Striped Bass (*Morone saxatilis*) is a game fish that lives in marine, estuarine and riverine environments during various stages of its life cycle. These three fish are predators, with the voracious Bluefish feeding on large quantities of Weakfish. In winter, they seek the offshore waters.

From November to April, Harbor Seals (*Phoca vitulina*) are frequently observed at Montauk Point State Park, about one mile west of the lighthouse. These marine mammals bask on the rocks at low tide and dine on fish, crustaceans and mollusks during the incoming and high tide. They can usually be seen in small groups.

Many northern species of shorebirds and waterfowl overwinter along Long Island's coastal waters. Common Eiders and White-winged, Black, and Surf Scoters are a few examples of waterfowl often seen in large numbers at Montauk Point State Park during the winter months.

The deeper ocean waters are home to large creatures of the nekton that include whales, migratory off the coast; sea turtles, wandering northward to Long Island from June to September; and squids, nocturnal mollusks found to a depth of up to 600 feet.

Terns are migratory diving birds frequently seen flying over the ocean and estuarine waters from May to September. The Common Tern (*Sterna hirundo*) is a colonial nester, scraping its small, shallow nest into sandy beaches in close proximity to the gulls' nests. Fish are the

Common Tern's main diet, and they are aggressive predators. In comparison, the Least Tern (*Sterna antillarum*) is a smaller, more passive bird that hovers in the air longer than the Common Tern before feeding on fish, small crustaceans, and insects. These terns prefer to lay their eggs away from the other tern and gull species. Do not go near the birds' nesting sites on the beach because, when disturbed, the terns will fly up in the air and out of the nest to ward off potential predators. When engaging in this behavior, terns leave their eggs unprotected against gulls, domestic cats, owls, foxes, raccoons and other actual predators.

The most abundant of the shorebirds are the gulls, not only found at all types of seashore environments, but also seen on Long Island in garbage dumps, strip mall parking lots, and occasionally beside garbage pails at suburban homes. The predominant species is the ubiquitous Herring Gull (*Larus argentatus*), an omnivorous scavenger which feeds on bird eggs, garbage, carrion, fish, invertebrates, or anything else it can find. Gulls drop hard-shelled invertebrates repeatedly from mid-air onto a hard surface until the shells break. *Larus* are colonial nesters, laying their eggs in a variety of locations.

Gulf Stream

Originating in the Gulf of Mexico, this large ocean current passes near Long Island's coast. Due to its close proximity, warm water and tropical species occasionally occur in Long Island's colder waters as these creatures are transported with the current. These species do not have a lasting presence in the local ecosystem, but it is important to note that unusual specimens may be found washed up on the shoreline or observed within our waters.

Barrier Island Dunelands

The barrier islands, a buffer of sand between the ocean and the mainland, are ecologically subdivided into the upper beach zone, the dune and swale zone, the thicket zone and sometimes the maritime forest. The width of the zones and the variety of species within these areas vary greatly, and the typical plant and animal inhabitants will be described in the following paragraphs.

Upper Beach Zone

The upper beach encompasses the area from the high tide line to the edge of the foredune. Beached Eelgrass leaves, seaweeds, *Spartina* fragments, crab exoskeletons, seashells, skate and whelk eggcases, and carrion are a few of the many components which comprise the beach wrack. This is a fundamental community of the upper beach as it provides a food source and shelter for many animals.

The most common inhabitant under the beach wrack is the Beach Flea (*Orchestia* spp.), a small, dark-colored amphipod crustacean. When you lift the wrack up from its position on the beach, many *Orchestia* can be seen jumping about. Beach Fleas are scavengers, feeding upon the decaying organic material found among the wrack. This amphipod provides a food source for shorebirds, such as Sanderlings, plovers, and turnstones.

Beach Pea, Saltwort, Dusty Miller, American Sea Rocket, Seabeach Orach and American Beach Grass are the plants that extend down to the uppermost portions of the upper beach from the primary dune, but none of these are a dominant component of the upper beach.

Dune/Swale Zone

Vegetation varies within the dune/swale zone as the swale is more protected and is subject to different environmental conditions than the dune. The dune/swale zone is subdivided into the foredune, the leeward side of the dune and the landward side of the swale.

The foredune is a severe environment with the challenges of high temperatures and winds, dry, sandy soils, salt spray and storm surges. The biota must be adapted to these conditions to survive in this location, and the plants discussed in this section meet the challenge.

An important coastal plant and a dominant component of the primary dune is American Beach Grass (*Ammophila breviligulata*), also called marram. Usually the first plant to grow on the dune, beach grass spreads by horizontal rhizomes close to the surface and attains water from deeply descending, fibrous roots. Its high tolerance to salt spray and wind help it survive on the foredune, and the plants' leaves and stems trap the blowing sand to help provide dune stabilization.

Once the dune has become established, other plants take up residence including Seaside Goldenrod, Dusty Miller, American Sea Rocket, Beach Plum, Beach Rose, Beach Heather, Poison Ivy, Prickly Pear Cactus, Virginia Creeper* and Bayberry.

On the leeward side of the dune there is more protection from the harshness of the environment with Bearberry* (*Arctostaphylos uva-ursi*) the dominant groundcover. Beach Heather (*Hudsonia tomentosa*), a low-growing, evergreen shrub, predominant in large, isolated patches, traps sand in its dense mats, contributing to dune growth and stability. Other plants on this portion of the dune include Beach Plum, Bayberry, American Holly, Seaside Goldenrod and the ubiquitous Poison Ivy.

The landward side of the swale establishes a gradual vegetation transition, with Bearberry*, Pitch Pine*, Earthstars*, American Holly, Black Cherry, Eastern Red Cedar, Highbush Blueberry* and Poison Ivy the typical plants found in this location.

Thicket Zone

The dense band of shrubs between the dune/swale zone and the maritime forest, called the thicket zone, varies in its width: sometimes a narrow ecotone, frequently a wide expanse of land. The thick vegetation provides shelter, either permanent or transitory, for many animals within the community. Plantlife located in this zone are Bayberry, Poison Ivy, Earthstars*, British Soldier Lichen*, Black Cherry, Pitch Pine* and American Holly.

*Illustrated in *A Field Guide to Long Island's Woodlands*

Maritime Forest

The swale behind the secondary dune and its crest give protection and stability to a unique barrier island community, the maritime forest. A layer of accumulated humus, consisting of decayed plant litter, provides the necessary nutrients to support this type of forest. The vegetation in this habitat is not fully tolerant of the saline conditions brought about from the salt spray in the ocean winds, resulting in a flat-topped forest canopy, and gnarled, thick trees that grow outward instead of upward.

One of the foremost examples of a maritime forest is found on Long Island at the Sunken Forest, Fire Island National Seashore. The principal canopy plants are American Holly (*Ilex opaca*), Sassafras* (*Sassafras albidum*) and Shadbush (*Amelanchier arborea*), a shrub that competes with the other two dwarfed trees. There are remnants of Pitch Pine, Eastern Red Cedar and oak from an earlier community that have been mostly replaced due to succession.

Because of the canopy's thickness there are few understory plants, but commonly found are Highbush Blueberry*, Inkberry*, False Solomon Seal* and young holly, Sassafras, and Shadbush saplings. Vines are plentiful and include Virginia Creeper*, Poison Ivy, Catbriar* and Fox Grape*.

There are several bogs and a freshwater marsh in the Sunken Forest supported by a lens of groundwater underneath Fire Island. Ferns, mosses, and cattails inhabit the bog, along with Tupelo* and Red Maple*; *Phragmites* is the dominant plant in the marsh.

*Illustrated in *A Field Guide to Long Island's Woodlands*

Animals of the Barrier Island Dunelands

Unlike the amphipods that remain under the beach wrack, the animals of the dunelands are mobile and may be encountered anywhere on the barrier island. The following species are commonly found in the dunelands:

Arthropods - Wolf Spiders*, Tiger Beetles, Sand Wasps, Velvet Ants, Sand Locusts, ants

Amphibians - Fowler's Toad*

Reptiles - Hognose Snake*, Black Snake*, Garter Snake*, Box Turtle*

Birds - Song Sparrow*, Mourning Dove, Gray Catbird*, Yellow Warbler, Brown Thrasher, Northern Mockingbird, Rufous-sided Towhee*, Northern Cardinal, American Robin*, Red-winged Blackbird*, Yellow-bellied Sapsucker, Fish Crow, Black-capped Chickadee, American Redstart, Carolina Wren, Common Yellowthroat

Mammals - Raccoon*, Red Fox*, White-footed Mouse*, Meadow Vole, White-tailed Deer*, Eastern Cottontail*

*Illustrated in *A Field Guide to Long Island's Woodlands*

CHAPTER III

Human Interaction

Historical Perspective

Native Americans

The shores of "fish-shape Paumanok", as Walt Whitman called Long Island, have been a treasured resource from prehistoric days to the present. The thirteen different groups of Algonquin Native Americans that settled here, possibly as early as 10,000 years ago, utilized the beaches much as we do today. They built their homes near the water to take advantage of the cooling breezes in summer and the abundant year-round supply of seafood.

Archaeological evidence indicates that the early inhabitants relied very heavily on clams for sustenance, although they enjoyed their fair share of crabs, oysters and scallops as well. Shellfish were easily gathered in the shallows by women and children, while the men fished or hunted wild game in the extensive forests inland. The empty shells, especially those of quahogs, were often used in surprisingly varied ways, from implements for digging, hoeing and scraping, to being ground up to temper clay pottery. One of the most notable uses was the making of beads or "wampum" from whelk and quahog shells. These beads (the purple ones from the clam having the greatest value) were used for personal decoration, for trade or barter, and for making belts which the tribal historian depended upon as a "memory device" to help recall specific events.

The Native Americans on Long Island became expert fishermen in shallow and very deep water, employing nets, hooks, spears and dugout canoes. Fashioned by burning and scraping out the interior of trees, the canoes, some 40' in length, were heavy and narrow with sloping ends. Propelled by wooden paddles, they were used not only for fishing but also for whaling. In a somewhat daring maneuver, men in several dugouts would intercept a whale off the outer beach and drive it ashore, where it was clubbed and speared by waiting villagers. Dugout canoes also provided the only means of transportation and were used for extensive travel, plying the bays, rivers, sound and ocean.

Later Inhabitants

The early European settlers followed the Native Americans' lead and established their villages along the shores, also using the sea as their primary method of communication and transportation. The many protected inlets around Long Island made travel and shipping by boat the most efficient and economical means, and small sailing vessels became a common sight. As the demand for ships grew, numerous towns along the coast became shipbuilding centers, spawning an industry that would last until the early 20th century. Prior to the American Revolution, Oyster Bay was the major shipbuilding community, although Southold and East Hampton made substantial contributions.

Whaling Industry

After the war, shipbuilding took a different tack, and construction of large whaling vessels became paramount. When the English colonists first arrived, they learned from the Native Americans about hunting whales near the shore with small boats. However, the number of whales declined so sharply that by 1800 it was virtually impossible to find them nearby, and men turned to pursuing them around the world in large ships. The lamps and machinery of the new Industrial Revolution created an enormous demand for whale oil, and Long Island soon took a lead in supplying the precious commodity. Until the 1860's, Sag Harbor and Greenport built and outfitted well over one hundred whaling ships, bringing tremendous growth to the East End of the Island. Farther west, Cold Spring Harbor grew into the second largest port, outfitting ships and providing employment for many, such as sail makers, blacksmiths, coopers and merchants. Crews were hired from the local population and often included Native Americans for their expertise. It was a time of prosperity for many, of world travel and exotic adventures. Many of Long Island's historic houses belonged to sea captains, and much of the local art and literature reflects the preoccupation with the "leviathans" of the deep.

Shipbuilding

The discovery of petroleum in the mid 1800's caused an almost overnight decline in the world's demand for whale oil, but shipyards across Long Island survived because of the constant need for large trading ships as well as small sailboats and fishing boats. Sloops and schooners plied the waters, carrying supplies to and from New York City and other coastal towns. The home ports for these fleets became the community centers where farmers, sailors and businessmen met. Port Jefferson, with its deep harbor and proximity to building materials, was a major hub until the introduction of steamships, although the successor of its first steam ferry to Bridgeport in 1883 is still running today.

Robert Fulton's invention of the steam engine revolutionized the shipping industry, greatly reducing travel time and increasing demand for use in commerce and commuting. Luxurious steamboats carried passengers between New York and New England, creating a new population: vacationers. The Brooklyn Navy Yard, which was once the largest industrial complex in the state, produced the first steam powered warship and the steam frigate which laid the first trans-Atlantic cable. For more than a century, it has also been the major shipyard servicing the steady flow of ships coming to New York Harbor from around the world. Nearby, in Greenpoint, Continental Iron Works launched the ironclad steam warship Monitor, which later fought the Merrimack to help save the Union fleet during the Civil War. The end of the 19th and the beginning of the 20th centuries saw the era of steamships, and no more so than on Long Island. But the introduction of the automobile and improved rail service caused a rapid decline, and the last steamers were requisitioned into service during World War II.

Fishing and Related Industries

Like the Native Americans, the earliest European settlers were dependent on the sea for food, and almost everyone fished to provide for their families. But fishing as an industry really began when the numerous boatyards on the Island provided the skiffs and other vessels necessary for grand-scale catches. Large fleets set out from Freeport and Greenport to supply the constant demand from New York City and local markets. The advent of steamships allowed the fishermen to go farther from shore, seeking more varieties and greater quantities. Fishing served as a way of life for many along both the North and South Shores and continues to do so to this day. Commercial and recreational fishermen ply Long Island Sound and the Atlantic Ocean, and their boats fill the harbors from Sheepshead Bay and Freeport to Captree and Montauk.

The economic well-being of countless Long Islanders came from the sea, in more ways than one. Small fish called Menhaden or Mossbunker were harvested for fertilizer, beginning in the 1790's when much of the Island's soil had become depleted from over-farming. Soon processing plants were built which manufactured not only fertilizer but also fish oil, and a booming national industry ensued. Long Island became the country's number one producer of both, with 23 industrial plants scattered along the South Shore and the Peconic Bay. The boom lasted until the 1930's, when petroleum became cheaper to process than fish oil and synthetic fertilizers were developed.

Shellfishing for clams, oysters and scallops, as well as lobstering, has been popular from the time of the earliest inhabitants to the present, but it wasn't until the mid-1800's that they became a huge industry on Long Island. This was when baymen started using steel net dredges to scrape off the bottom, which greatly increased their "haul." Numerous oyster packing houses dotted the South Shore and bays on the East End, as well as Huntington and Smithtown, supplying the growing demand in both New York City and Europe. By the turn of the century, oystering was a multimillion dollar business, with enormous fleets employing thousands. Baymen "seeded" the beds in an effort to insure a continual supply, but eventually overharvesting did take its toll. The biggest blow to the industry, however, was the Hurricane of 1938 which created a new inlet into the Great South Bay, increasing the salinity and killing the oysters. Clams, which are more tolerant to variations in salinity, became the dominant species and therefore the new "crop". Although they were not as "fashionable" or in demand, many were sold for chowder or for eating fried or raw. Scallops, especially those from the Peconic Bay, have long been a treasured treat to Long Islanders, but their numbers were never as great as those of oysters and clams. Nevertheless, all three have been greatly reduced by pollution and over-harvesting. And even the lobster industry, which for over a century rivaled Maine, has recently fallen into hard times as a result of a mysterious disease which has destroyed most of Long Island's lobsters. The once lucrative, albeit strenuous, life of baymen and other fishermen has become increasingly difficult.

Salt Grass

Another industry that flourished for a period on Long Island's shores was the harvesting of salt grass from the numerous marshes that ring our bays. "Marshing" or gathering the long grasses in the fall, became popular before the Revolutionary War when the demand for cattle feed grew, and it reached its peak in the mid-1800's. A decline in livestock on the island and the introduction of vitamin-enriched grains caused its demise, although landscapers still often seek "salt hay" as a nutrient-rich mulch.

Today's Issues

As human population and industrial development increased in Nassau and Suffolk Counties, the five boroughs, and the surrounding regions, the physical and ecological dynamics of Long Island's waters and shorelines were negatively affected. Pollution, hypoxia, runoff, erosion and sedimentation, dredging, the filling in of wetlands, over-fishing, and the introduction of alien species were all conditions brought upon by this growth. These problems were mainly ignored and left unchecked until 1970 when legislation began to be enacted to remedy many of these concerns. Ongoing programs continue the cleanup and only the full cooperation between government, industry, and the general public will help alleviate the problems.

Non-point Source and Point Source Pollution

Pollutants entering our waters and shorelines are defined by their origins. Point source pollution has a specific source such as a sewage treatment or an industrial plant. Non-point source pollution (NPS) is carried by runoff, also known as stormwater, and does not have a specific source. It is transported by rainfall or snowmelt that moves over and through the ground. Sources of NPS pollution are agriculture and urban runoff, faulty septic systems, recreational boating, and changes to the environment, such as wetland destruction or storm channel paths. Runoff can carry natural and manmade pollutants including fertilizers, herbicides, and insecticides, oil, grease, and toxic chemicals, sediment from erosion, salt, nutrients such as nitrogen, floatable debris, pathogens, heavy metals and airborne pollutants. Pollution can harm wildlife, kill native vegetation, foul drinking water, and close recreational areas.

Buildings, roads, parking lots, and bridges are composed of non-porous materials that do not allow runoff to slowly percolate into the ground. Water sits on the surface, accumulates and then runs off in large quantities. Storm sewers are installed in cities to channel this urban runoff which gathers speed once it enters the sewer system. Large volumes of runoff eventually empty into a location such as a stream causing erosion, damaging the vegetation and widening the channel. Native aquatic life cannot survive well under such conditions.

Pollutants

Excessive amounts of nutrients, such as nitrogen, can have diverse effects on the plant and animal life in a marine ecosystem. Increased concentrations of nitrogen in the aquatic environment cause huge algae blooms, leading to an abundance of decaying algae. This deteriorating matter causes lower concentrations of dissolved oxygen. If the amount of dissolved oxygen falls below 1.5 mg/l at any time or below 3.5 mg/l for a short period, i.e. less than a week, hypoxia results. Hypoxia has caused fish kills in Long Island Sound, Hempstead Harbor and other locations on Long Island. Depleted oxygen levels also reduce the adult fish population, retard the growth of juvenile lobsters and flounder, cause lobsters and other species to be less resistant to disease, and kill slow-moving species such as lobsters, starfish, menhaden and sea robins.

Disease carrying viruses, bacteria, and other microorganisms are collectively known as pathogens. Pathogens are borne in our waters from sewer overflows, malfunctioning septic tanks and sewage treatment plants, and runoff containing contaminated human or animal waste. There are too many individual tests required to check for each disease in our waters, so scientists test for coliform, a harmless bacteria found in animal and sewage waste. If the coliform levels are high, beaches are closed and shellfishing is stopped until repeated tests in the area return to within the acceptable range.

Toxic chemicals are defined as substances that can kill or damage a living creature. Synthetic organic chemicals and heavy metals are examples of toxins, most being carcinogenic. Heavy metals, such as copper, lead, mercury, cadmium, nickel, chromium and zinc, are slightly soluble in water. Harmful in low concentrations, they accumulate in animal flesh when ingested and in sediment on the bottom of the oceans and bays. There has been an overall decrease of these in the environment since the 1970's due to the shutdown of older factories and the Clean Water Act.

Organic substances that are toxic chemicals include polycyclic aromatic hydrocarbons (PAHs), polychlorinated biphenyls (PCBs), pesticides, and petroleum. PAHs are present in coal and petroleum and form when fuel is burned and when petroleum spills into water from barges and industrial business. PCBs, used for electrical transformer insulators, are no longer manufactured in the United States. They do not break down in the environment, so they are still a hazard. Pesticides are defined as any organic compound used to kill plant or animal life. The first three organic substances mentioned above are carcinogens, with PCBs and pesticides also causing birth defects.

Petroleum, unlike the other organic substances, does break down or evaporate over time. Its short-term effects are very damaging and prove lethal for plankton, fish larvae, aquatic birds, and shellfish. It is spilled from oil barges into our waters as a point source pollutant. Usually the spill is small, but occasionally a huge spill occurs, having a devastating effect on the ecosystem.

Debris that floats in coastal waters and washes up on the beach is known as floatables. Some examples of this are plastic, glass, cigarette filters, and aluminum beverage cans. The largest source of this garbage is combined sewer overflows and storm drain waters, but its origins can be from tributary waters, boaters and beach visitors. Marine organisms can ingest or become entangled in the floatables, often resulting in death. The debris can foul propellers and damage engines of boats, and spoil the beauty of our shorelines. In 1988, there was a surge of beach debris, including medical waste. A severe economic loss occurred for our region as beach attendance dropped, and there was a decline in business at seafood restaurants.

Airborne pollutants, primarily nitrogen oxide and sulphur oxide, can also affect our waters. They combine with water in the atmosphere and form acid rain, harming the ecosystem. Their point source origin is industrial pollution and fuel combustion.

Erosion, the wearing away of soil and other sediment by water, can be a pollutant to the ecosystems of aquatic environments. Stormwater runoff, land cleared by farmers or developers, or any other disturbance of the land and its vegetation can result in erosion. The sediment can remain suspended for a period of time, blocking out sunlight and impeding biological activity. A long-term effect of sedimentation in shallow coastal regions, such as wetlands and embayments, is that the silt can wipe out fish spawning areas and other marine life.

Wetland Destruction

Wetlands, such as salt marshes, are watersheds (drainage basins) that act as a link between land and water resources. They are breeding grounds for many species and help filter pollutants from land runoff, including nutrients. Although the wetland area can filter pollutants out of the water, too many pollutants in the runoff will kill the biota living in the salt marsh. Besides the effects of erosion and sedimentation on the wetlands previously discussed, there are additional forms of human disturbance that can lead to the destruction of a wetland. Marshes often have been drained and filled for land development, agriculture and industry, and have served as a dumping ground for dredged material. They have been disturbed and restricted by ditching for mosquito control, drained for flood control and the production of salt hay, diked to create impoundment (flooded areas) and impeded by bulkheads and jetties at adjacent, neighboring coastal areas. Wetland protection laws were passed in the 1970s, but some areas already damaged from previous development have become degraded.

Invasive Non-native Species

An invasive non-native species is an organism that is intentionally or accidentally introduced to a region. The natural controls of their native ecosystem may be nonexistent in their new habitat, so they crowd out the native plants or animals to become the dominant species. Two examples, the Zebra Mussel and the Gypsy Moth, have caused economic and ecological losses in freshwater bodies and forests respectively. *Phragmites australis*, the Common Reed, is an example of an invasive non-native species found in our Long Island wetlands. Growing in disturbed areas, it displaces the local plants of this fragile ecosystem. An example of an introduced animal is the Common Periwinkle, *Littorina littorea*. Originally from Europe, it floated to the East Coast of the United States during its free-swimming larval stage. Once established in Nova Scotia, it migrated south, eventually extending its range throughout the entire eastern seaboard. Though not an economic threat, it dominates other native species within its habitat.

Dredging

Dredging, the deepening of channels to maintain adequate depth for ships and boats, is a needed endeavor for safe navigation. New York's major harbors are dredged of bottom sediment to allow vessels to enter the harbor for commerce and industry. However, most of this dredged material contains pollutants that have accumulated on the ocean bottom for decades. The majority of this material gets dumped in the open ocean, affecting the ecosystems of those dumping grounds and contaminating those regions. These contaminants can be transmitted through the food web and eventually reach humans. Locations used for dumping have been off limits to fishing and shellfishing due to the high levels of toxins. Modernization of dredging methods is required to minimize the volume of amassed materials and an efficient method of testing and treating this contaminated material would allow some of it to be used for fill or beach nourishment.

Overfishing

Any marine animals that are harvested commercially, such as fish, lobsters, crabs, scallops, clams etc., can become the victims of overfishing. The adult population can be fished so heavily that it is unable to replenish itself through reproduction, and the species is extirpated from the ecosystem. Overfishing occurs when the animals are harvested at a smaller than standard size (before they reach maturity) or when too many specimens are harvested from a specific area in a short period of time. Species size and quantity limits are regulated for recreational and commercial fishing. Licenses, permits and many strict regulations are imposed on commercial fishing vessels to prevent the overfishing of their intended or unintended catch.

Correcting the Problems

Public awareness and concern over the past thirty years have produced an abundance of government agencies, private organizations, and federal and state legislation that addresses our environmental concerns. These support the programs, funds, and grants that provide the research, cleanup, enforcement, education, and prevention needed to correct the current problems. Each has a specific mission in the conservation of our environment with some focused particularly on Long Island's issues. It is impossible to list all of the groups and laws in this book, but several are cited as examples.

A few of the many federal agencies include the U.S. Environmental Protection Agency (EPA), the U.S. Fish and Wildlife Service, and the National Marine Fisheries Service, a division of the National Oceanic and Atmospheric Administration (NOAA). They help enforce and enact such funds and laws as the Superfund, the Clean Air Act, the Clean Water Act and the Coastal Zone Management Act.

Locally, the Long Island Sound Study, a program designed to report on the issues affecting the Long Island Sound, is sponsored by the EPA along with the states of New York and Connecticut. Some of the state agencies are the New York State Department of Environmental Conservation (NYSDEC), Sea Grant and the New York/New Jersey Harbor Estuary Program.

The National Audubon Society, The Nature Conservancy, the Sierra Club, Greenpeace, and the National Wildlife Federation are a small sampling of the many private national organizations available to the general public. The first three previously listed have local chapters in our area. Private organizations that focus solely on the conservation of Long Island's natural resources include Save the Peconic Bays, Group for the South Fork, Save the Sound and Save Our Shores.

You Can Make A Difference

An effective way to help with the preservation of Long Island's waters is to share your opinions on its conservation and become involved. Attend public meetings that discuss the decisions being made about environmental programs within the community. Write letters to members of our government to state your position on an environmental issue. Volunteer for such tasks as beach cleanups or recycle days sponsored in your area. Join any of the federal, state, or local organizations that support conservation and share your views.

Save and properly dispose of items that are commonly accepted as recyclable as it will not only allow the reuse of the natural resources, but will minimize the amount of floatable debris found on our shores. Most communities on Long Island have specific pick up days for these materials which include plastic containers marked Code 1 and 2, glass, steel, aluminum cans and foil. Scrap aluminum is generally accepted but not most other metals. Most paper such as newspapers and white paper can also be recycled. Usually another specific day of the week will be set aside for pick up of "paper only" in your community. Corrugated cardboard and phone books are not always accepted and items such as waxed paper, milk cartons, paper napkins, tissues, fax paper and stickers cannot be recycled. It is important to set aside only the accepted materials for recycling because unaccepted items may contaminate the reclamation process. Check with your garbage carriers to find out what is and is not acceptable for recycling in your community.

Automotive batteries, rechargeable batteries, paints, turpentine, solvents, and other household toxics should not be placed in the regular trash or washed down drains. Check with your local recycling agency to find out where and when you can drop off these items.

Certain materials and substances require proper handling to avoid contaminating runoff. One example, motor oil, is classified as a hazardous waste as it contains heavy metals and other toxic substances. An automobile leaking oil from its engine onto the street can contaminate the groundwater and contribute pollution to the runoff. Proper car maintenance is a simple solution to this problem. If you change the motor oil in a car never dump it down a storm drain as the oil will flow untreated into our waters. Save the oil in a plastic milk jug and drop it off at either a local recycling agency or a quick-lube shop that will handle the used motor oil.

Animal waste is another common substance that decomposes into the runoff, possibly contributing dangerous pathogens. Scoop up pet waste, and if it is not mixed with cat litter, flush it down the toilet. Do not drop waste directly into storm drains. If your community does not already have them, find out about adopting laws for pet waste stations at convenient locations and enacting requirements for pet owners to pick up their animals' waste.

Fertilizers, used on our lawns and gardens, are a third example of materials that need to be used properly and sparingly to avoid entering the runoff. Too much nitrogen in our bays and waters can cause hypoxia or brown tide, both detrimental to marine plants and animals. Some precautionary measures include not fertilizing before a rain storm, avoiding the placement of materials on sidewalks and driveways, testing your soil, and using compost, a natural fertilizer. Leave your grass clippings on your lawn to act as a natural mulch; this also reduces trash volume.

Composting, another means to reduce refuse volume, is a process whereby organic plant and animal materials, through a microbial process, are converted into a usable natural fertilizer. Leaves are the dominant component used in composting, along with grass clippings, twigs, branches, straw, vegetable and flower garden refuse, bone meal and others. Composting is practiced in many town and county facilities, although many homeowners find it economical and convenient to make their own compost. When you are interested in setting up a compost pile of your own, instructions and products are available in many garden catalogues.

Suggestions for other things you can do to protect our environment include purchasing recycled products, reducing the volume of packaging you buy, and reusing or donating items such as laser/ink printer cartridges, computers and eyeglasses. In addition, at the grocery store, use your own canvas bags, request paper instead of plastic bags, or refrain from using any bag at all when buying just a couple of items. Do not litter; it increases the volume of floatable debris found at the beaches or on the water and may be harmful to marine life.

Plant Species

Beach
Rose

MTW

TREES - Tall, woody plants with single or divided trunk and numerous branches, minimum adult height 15-20 feet; deciduous or evergreen.

Eastern Red Cedar (*Juniperus virginiana*)
Evergreen   Height: up to 60 ft.   Dioecious
Most common eastern conifer; bark reddish-brown, thin, easily shredded. Young leaves prickly, tiny, dark green needles, sharply 3-sided, in pairs; softer, more scale-like at maturity. Cones waxy, dark blue, berrylike. Heartwood aromatic, made into cedar chests and strips to prevent insect infestation.
Habitat: Maritime forest, salt marsh edge, upper beach, dunes, swale

Black Cherry (*Prunus serotina*)
Deciduous   Height: up to 80 ft.   Monoecious
Largest cherry tree; bark distinguished by horizontal markings, smooth and reddish-brown in young trees, rough and dark at maturity. Leaves 2"-5", alternate, elliptical, finely saw-toothed, tapered tip, often hairy underneath, turn yellow or red in fall. Crushed leaves and bark have distinctive, unpleasant odor and bitter taste. Flowers small, white, in slender, terminal raceme. Fruit edible drupe, dark purple, June-October.
Habitat: Maritime forest, thicket zone, dunes, swale

American Holly (*Ilex opaca*)
Evergreen   Height: up to 50 ft.   Dioecious
Slow-growing tree, readily identifiable by its sharp-pointed, spiny, thick, leathery leaves, 2"-4" long. Bark yellowish-gray, either smooth or rough and lumpy, often developing large stubby projections on trunk. Flowers small, white, on male and female trees. Fruit, on female trees, berrylike, bright red in autumn, usually lasting through winter. Used ornamentally at Christmas.
Habitat: Maritime forest, thicket zone, dunes, swale

Post Oak (*Quercus stellata*)
Deciduous   Height: up to 50 ft.   Monoecious
In White Oak Group; bark gray, rough, furrowed. Leaves 4"-8", alternate, shiny dark green, turning brown in fall, hairy underneath, leathery, 5-lobed shape resembles a Maltese cross. Flowers yellowish-green catkins. Fruit elliptical acorn with bowl-shaped cup, matures in one season.
Habitat: Maritime forest

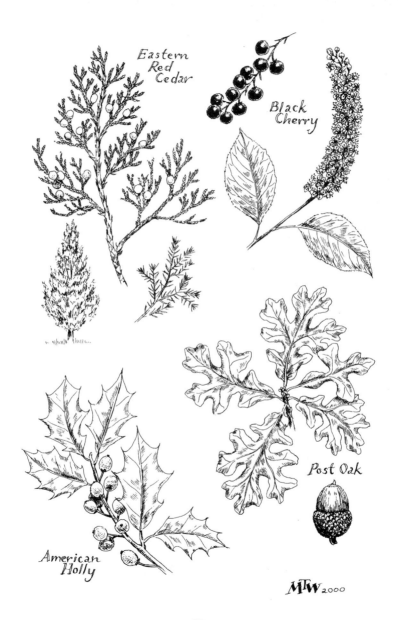

Eastern
Red
Cedar

Black
Cherry

Post Oak

American
Holly

MTW 2000

SHRUBS - Woody plants, less than 15-20 ft. in height, normally with several trunks and numerous branches.

WILDFLOWERS - Blooming herbaceous plants, sometimes annual but most often perennial.

Bayberry (*Myrica pensylvanica*)
Shrub   Height: up to 8 ft.   Flowers: Spring to early summer
Dioecious; bark smooth, gray; twigs often hairy. Leaves 1"-5", oval, serrate, pointed tip, dark green, shiny above, hairy below, fragrant. Flower small ament. Fruit round, grayish-white, waxy berry in clusters, used to make candles since colonial days.
Habitat: Maritime forest, thicket zone, dunes, swale

Prickly Pear Cactus (*Opuntia compressa*)
Wildflower Height: up to 10 in.   Flowers: Late spring-early summer
Only cactus indigenous to Long Island. Jointed, branched, ovoid stems, mostly prostrate, fleshy to retain water, with sharp spines and clusters of minute brown bristles. Flower bright yellow, 2"-3", waxy. Fruit pulpy, reddish-purple, 2" long, edible.
Habitat: Upper beach, dunes, swale

Dusty Miller (*Artemisia stelleriana*)
Wildflower   Height: up to 2 ft.   Flowers: Late spring to summer
Common mat-forming plant, introduced from northeastern Asia. Leaves silvery, multi-lobed, tomentose for protection from salt spray and dehydration. Flowers small, yellow, rayless composite, on terminal spike. Used ornamentally in gardens.
Habitat: Upper beach, dunes, swale

Woody Glasswort (*Salicornia virginica*)
Wildflower   Height: up to 9 in.   Flowers: Late summer to fall
Perennial halophyte. Succulent spikes grow from central woody stem. Leaves tiny scales, opposite, at stem joints. Flower minute, green, inconspicuous, in leaf axils. Plant turns gray in autumn. Edible, salty, often used in salads and pickled. **Slender Glasswort** (*S. europaea*) an annual to 1 ft., similar but with mostly erect stems, multi-branched; plant spreading, green, turning red in autumn (not illustrated).
Habitat: Salt marsh

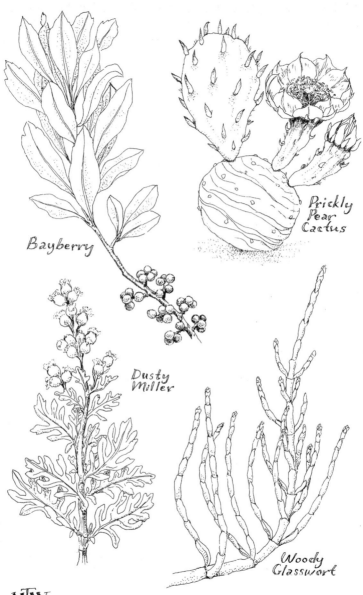

Bayberry

Prickly
Pear
Cactus

Dusty
Miller

Woody
Glasswort

MTW2000

Seaside Goldenrod (*Solidago sempervirens*)
Wildflower   Height: up to 4 ft.   Flowers: Summer to fall
Abundant, tolerant to a variety of ecological conditions. Stem stout and
erect. Leaves fleshy, lance-shaped, up to 1', sessile, sometimes clasping, toothless. Flowers bright yellow, in thick plume-like clusters on
upper portion of stem. Fibrous roots help stabilize dunes.
Habitat: Upper salt marsh, dunes, swale, upper beach

Groundsel Bush (*Baccharis halimifolia*)
Shrub   Height: up to 10 ft.   Flowers: Late summer to fall
Dioecious; fruits form conspicuous fluffy white clusters in autumn.
Stem woody, branching. Leaves up to 3", alternate, thick, coarsely
toothed on lower stem, otherwise toothless. Flowers tiny, yellowish-
white clusters. Fruit on pistillate plants.
Habitat: Upper edge of salt marsh

Beach Heather (*Hudsonia tomentosa*)
Wildflower   Height: up to 8 in.   Flowers: Spring to mid-summer
Evergreen, tufted, spreading, multi-branched mats. Leaves minute,
scale-like, tomentose, greenish-gray. Flowers tiny, 5-petaled, yellow,
in clusters. Fruit small capsules.
Habitat: Upper beach, dunes, swale

Poison Ivy (*Toxicodendron radicans* = *Rhus radicans*)
Shrub/vine   Height: variable   Flowers: Late spring to mid-summer
All parts, including stems and roots, toxic year round. Grows as upright
shrub, climbing vine, or trailing ground cover. Clings by aerial rootlets.
Compound leaves, 3 leaflets, usually shiny, long-stalked, turning red or
yellow in fall. Flowers inconspicuous, greenish. Fruit small, whitish,
berrylike, in clusters; important food for birds. Helps stabilize dunes.
Habitat: Everywhere above high tide line

Marsh Mallow or Swamp Rose Mallow (*Hibiscus moscheutos*)
Wildflower   Height: up to 6 ft.   Flowers: Summer
Indigenous Long Island hibiscus. Numerous tall stems from shallow,
perennial root. Green, toothed, heart-shaped leaves up to 7" long,
alternate, underside hairy. Flowers large, up to 7" wide, pink, showy,
5-petaled, conspicuous column of stamens in center. Fruit ovoid, 5-
valved capsule, 1", seeds hairy, kidney-shaped; remains on stem
through winter.
Habitat: Brackish marshes

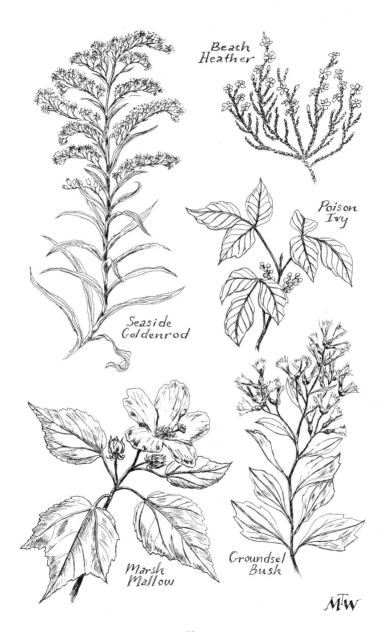

Beach
Heather

Poison
Ivy

Seaside
Goldenrod

Marsh
Mallow

Groundsel
Bush

MTW

Marsh-elder (*Iva frutescens*)
Shrub   Height: up to 8 ft.   Flowers: Summer to early fall
Dominant plant along upper edge of salt marsh. Superficially resembles Groundsel Bush. Leaves ovate, opposite, fleshy, serrate, up to 4". Flowers small, greenish-white, in terminal clusters. Fruit small ovoid achenes.
Habitat: Salt marsh edge

Beach Pea (*Lathyrus japonicus*)
Wildflower/Vine   Length: up to 3 ft.   Flowers: Late spring to summer
Legume; bacteria on root nodules supply nitrates to plant. Stems creep along ground, or climb up other plants. Leaves pinnately compound, 6-12 oblong leaflets, branching and curling tendrils at tip of leafstalk; triangular stipules at base. Flowers pea-like, in clusters, lavender to violet. Fruit elongate pod with edible peas, green, turning brown at maturity.
Habitat: Upper beach, dunes

Beach Plum (*Prunus maritima*)
Shrub   Height: up to 6 ft.   Flowers: Spring
Dense, with numerous branches. Leaves alternate, ovate, finely toothed, 2", pubescent beneath. Flowers white, profuse, in clusters on sides of twigs, precede leaves, turn pinkish before falling. Fruit edible drupe, purple, globose, 1/2" diameter, ripens in autumn. Used for making jelly.
Habitat: Upper beach, dunes, swale

Beach Rose or Salt-spray Rose (*Rosa rugosa*)
Shrub   Height: up to 6 ft.   Flowers: Mid-spring to summer
Introduced from eastern Asia. Branches numerous, thick, arched, very thorny. Leaves dark green, thick, wrinkled, pinnately compound, 5-9 ovate, toothed leaflets. Flowers 4" diameter, 5 petals deep pink, sometimes white, yellow center. Edible fruit 1", smooth, round, bright red hip, long sepals.
Habitat: Upper beach, dunes, swale

Saltwort (*Salsola kali*)
Wildflower   Height: up to 2 ft.   Flowers: Summer
Annual. Tall, erect, very prickly, bushy, multi-branched herb. Stems often with purplish-red vertical lines. Leaves succulent, pointed, grayish-green, barbed at tip. Flowers small, whitish, solitary, at leaf axils. Seed embryo coiled into conic spiral.
Habitat: Upper beach, lower edge of foredune

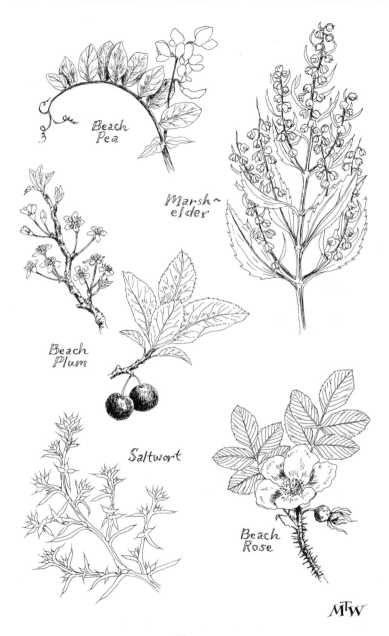

Beach
Pea

Marsh~
elder

Beach
Plum

Saltwort

Beach
Rose

M̄T̄W

Sea Blite (*Suaeda maritima*)
Wildflower   Height: up to 18 in.   Flowers: Summer to fall
Annual, introduced from Europe. Multiple ascending branches from
decumbent stems. Leaves pale green, often whitened, alternate, fleshy,
entire. Flowers small, greenish, at leaf axils. Fruit tiny, round, shiny,
dark red-brown.
Habitat: Salt marsh, upper beach

Sea Lavender (*Limonium carolinianum*)
Wildflower   Height: up to 2 ft.   Flowers: Summer to early fall
Multi-branched, shrublike, found in dense colonies, forming colorful
sprays in fall. Leaves basal, flat, leathery, oblong, 6", prominent
midvein. Flowers numerous, tiny, 5 petals, pinkish-purple, fragrant,
solitary, along one side of stem. Fruit one-seeded with mature ovary.
Habitat: Salt marsh

Seabeach Orach or Marsh Orach (*Atriplex patula*)
Wildflower   Height: up to 2 ft.   Flowers: Summer to fall
Highly diverse annual. Leaves fleshy, variable (usually arrow-shaped),
light green, mostly alternate but some opposite. Flowers small, green,
in panicles on leafless spikes from upper leaf axils. Fruit tiny seeds in
bract.
Habitat: Salt marsh, upper beach

American Sea Rocket (*Cakile edentula*)
Wildflower   Height: up to 1 ft.   Flowers: Summer
Low, creeping, branching succulent seaside mustard. Leaves fleshy,
alternate, oblong, toothed or lobed, narrower at base, 3"-5". Flowers
small, pale purple, 4 petals, in terminal clusters. Fruit green, 2-celled
rocket-shaped pod with one or two seeds; ripe capsule drops into sand
and sprouts the following spring. Extremely tolerant of saline conditions. Edible leaves bitter like horseradish.
Habitat: Upper beach, dunes

Shadbush, Juneberry or Serviceberry (*Amelanchier arborea*)
Shrub   Height: up to 25 ft.   Flowers: Early to mid-spring
Tall, often tree size (rarely 40'-60') with light gray bark. Leaves
alternate, oval, pointed, finely toothed, pubescent beneath, turning
yellow to red in fall. Flowers before leaves, white, 5 narrow petals, long
stalked in drooping terminal clusters. Fruit purplish, 1/2" drupe, edible
but dry, ripens June to July.
Habitat: Maritime forest

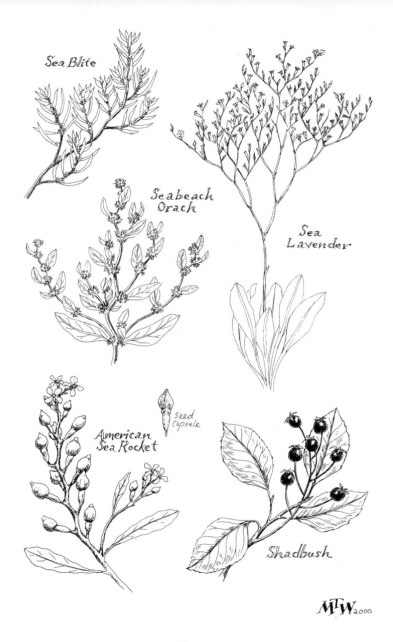

Sea Blite

Sea Lavender

Seabeach Orach

Sea Lavender

American Sea Rocket

seed capsule

Shadbush

MTW 2000

AQUATIC MARINE PLANTS - Rooted, submerged; produce seeds.

Eelgrass (*Zostera marina*)

Height: up to 4 ft.   Flowers: Summer
Habitat for many marine animals and plants. Stems narrow, branching, arise from creeping runner, spreads by rhizome. Leaves ribbonlike, alternate, multiveined, up to 3' long, 1/2" wide. Flowers small, greenish, concealed in sheath on one side of leaf. Fruit cylindrical seed. Extremely abundant in the Great South Bay.
Habitat: Bays, estuaries; sand; subtidal

GRASSES - Plants with round or flattened, jointed, hollow stems; split leaf sheath.

Saltmarsh Cordgrass (*Spartina alterniflora*)

Height: variable, up to 6 ft.   Flowers: Late summer
Dominant plant in lower salt marsh areas; large colonies. Stem short, round, hollow, spongy at base. Leaves flat, elongate, smooth, up to 2' long, 1/2" wide, tapered to point. Flowers minute, greenish, in erect panicle composed of spikes and spikelets; turn beige upon maturity. Fruit small seed grains. Roots and rhizomes bind sediment to form a spongy layer of grayish brown salt peat.
Habitat: Salt marsh

Saltmeadow Cordgrass (*Spartina patens*)

Height: up to 2 ft.   Flowers: Summer to fall
Indicative plant of higher salt marsh areas; forms dense flattened mats. Stem thin, hollow, recumbent. Leaves very thin, long, margins rolled inward. Flowers erect panicle, typically with 3-4 greenish spikes, browning at maturity. Fruit tiny seed grains. Roots and rhizomes bind sediment.
Habitat: Salt marsh

American Beach Grass (*Ammophila breviligulata*)

Height: up to 4 ft.   Flowers: Late summer
Most common plant on dunes. Stem stout, rigid, tall, arising from branching, spreading rhizome. Leaves long, narrow, arching. Flowers spiked panicle, whitish, turning light brown upon maturity. Fruit small seed grains. Roots and rhizomes stabilize dunes by binding sand.
Habitat: Upper beach, dunes, swale

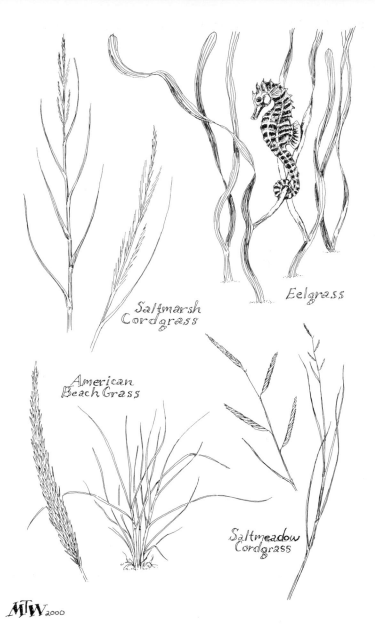

Saltmarsh
Cordgrass

Eelgrass

American
Beach Grass

Saltmeadow
Cordgrass

MTW 2000

Common Reed (*Phragmites australis*)
Height: up to 15 ft.   Flowers: Late summer
Tallest non-woody plant in wetlands; abundant, invasive, forming
dense stands. Leaves bladelike, up to 20" long, 2" wide, smooth, stiff,
sheaths loose and overlapping. Flowers purplish plume-like panicle.
Fruit seeds in silky-bearded feathery terminal cluster. Spreads by long
rhizomes and stolons. Indicative of disturbed natural environments.
Habitat: Salt marsh, brackish wetlands, freshwater marsh

Spike-grass (*Distichlis spicata*)
Height: up to 2 ft.   Flowers: Summer
Dioecious; often interspersed in matted stands with *Spartina patens*.
Stem erect, rigid, hollow, growing from rhizomes. Leaves numerous,
alternate, thin, rolled inward, overlapping sheaths. Flowers whitish-
green in thick, spikelike panicle; male spikelet longer than female. Fruit
tiny, white grains.
Habitat: Salt marsh

RUSHES - grasslike; leaves and stems often hollow; fruit seeds in capsule.

Black Grass (*Juncus gerardii*)
Height: up to 1.5 ft.   Flowers: Summer
Dense clusters. Stem erect, creeping roots. Leaves flat, elongate, round
in cross-section, loosely clasping sheaths; basal and 1 or 2 along stem.
Flowers on side of stem in terminal panicle. Fruit dark brown capsule,
purplish ribbing.
Habitat: Salt marsh (between marsh and drier upland), brackish shores

SEAWEEDS - Marine algae, most with stalks, fronds, attached to
substrate by holdfast; no vascular system, nourishment directly from
seawater. Three phyla: green (Chlorophyta), brown (Phaeophyta), red
(Rhodophyta).

Dulse (*Rhodymenia palmata*)
Rhodophyta   Length: up to 15 in.
Usually in swaying beds. Stalk tapered, forked, marginal leaflets.
Blade flat, thick, ribless, leathery, irregularly divided, reddish-purple,
opaque. Edible.
Habitat: Ocean, bays, estuaries; on rocks; mid-intertidal to deep subtidal

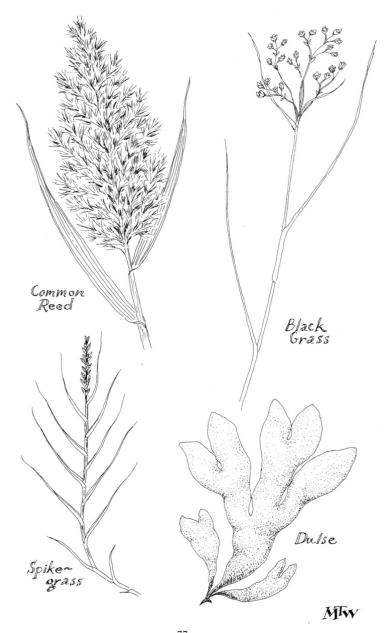

Common
Reed

Black
Grass

Spike-
grass

Dulse

MTW

Green Fleece (*Codium fragile*)

Chlorophyta   Length: up to 3 ft.

Introduced from Europe in 1950's, now widespread. Erect, dark green, bushy, spongy; thick, Y-shaped branches, velvety in appearance. Bleaches to white on shore. Often attaches to shells and rocks. Where abundant, has a choking effect on shellfish. Large masses resembling rope wash ashore. Edible.

Habitat: Bays, estuaries, ocean; on rocks, shells; intertidal to subtidal

Southern Kelp (*Laminaria agardhii*)

Phaeophyta   Length: up to 10 ft.

Relatively common. Only kelp in Long Island's waters, primarily occurring offshore. Sometimes found on beaches after storms. Extremely long-bladed, shiny brown, rubbery (thin, ruffled edges like lasagna noodles in summer; thick, straight edges in winter), lacking midrib. Short stem; multi-branching holdfast.

Habitat: Bays, estuaries, ocean; on rocks; intertidal to subtidal

Irish Moss (*Chondrus crispus*)

Rhodophyta   Length: up to 10 in.

Very common, often washing on beaches in large amounts. Leathery, variable in shape; color primarily purplish-red, occasionally brown or green, bleaches to yellowish; often iridescent blue under water. Blade ribless, fan-like, flattened, multi-forked, sometimes ruffled in appearance. Short stalk. Carrageenin, made from this species, used commercially as a gel in food, medicine and cosmetics.

Habitat: Bays, estuaries, ocean; on rocks, shells; intertidal to subtidal

Bladder Rockweed (*Fucus vesiculosus*)

Phaeophyta   Length: up 3 ft.

Abundant, dense masses serve as protection for crabs and other fauna. A host for epiphytes and epizoans. Wavy-edged olive-green flattened leathery blade with distinct midrib bifurcates repeatedly and contains paired air bladders. Attaches to rocks and shells in intertidal zone, frequently in association with Knotted Wrack. Also grows around bases of salt marsh grasses and pilings. Fleshy receptacles covered with numerous small protuberances.

Habitat: Bays, estuaries, ocean, salt marsh; on rocks, shells; intertidal

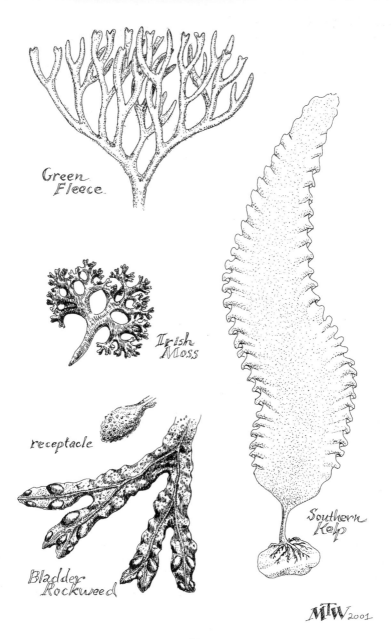

Green
Fleece.

Irish
Moss

receptacle

Bladder
Rockweed

Southern
Kelp

MTW 2001

79

Sea Lettuce (*Ulva lactuca*)

Chlorophyta   Length: up to 2 ft.

Initially attached by short stalk, eventually free-floating, often washing up in abundance on shorelines and mudflats at low tide. Blade thin, translucent, bright green with plain or ruffled edges. Thrives in nutrient-enriched water. Edible.

Habitat: Ocean, bays, estuaries; mudflats, rocky shores; intertidal to subtidal

Green Seaweed (*Enteromorpha intestinalis*)

Chlorophyta   Length: up to 2 ft.

Also known as Link Confetti. Tubular bright green unbranched fronds, sometimes with trapped air bubbles, occasionally free-floating. Wide tolerance of salinity and nutrient levels. Important food for fish and invertebrtates.

Habitat: Bays, estuaries, ocean; on rocks, shells; intertidal to subtidal

Knotted Wrack (*Ascophyllum nodosum*)

Phaeophyta   Length: up to 2 ft.

Brown-green fleshy narrow blades with no midrib contain single air bladders and forked branchlets. Stalked, yellow, warty receptacles in early spring. Sometimes adrift as oceanic plankton. Packing material for bait worms, lobsters.

Habitat: Bays, estuaries, ocean; rocky shores; intertidal

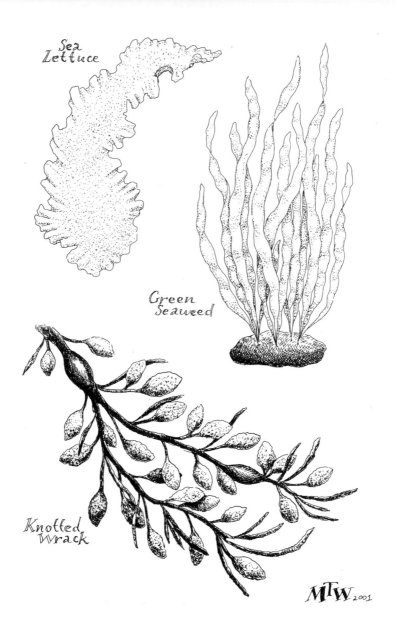

Sea
Lettuce

Green
Seaweed

Knotted
Wrack

MTW 2001

81

Animal Species

Piping Plover

MTW 2001

INVERTEBRATES - Animals that lack a backbone.

SPONGES - Simplest multicellular animals; porous bodies; exoskeletons comprised of glassy or calcareous spicules; mainly marine.

Red Beard Sponge (*Microciona prolifera*)
Height: up to 8 in.
Very common, especially in Long Island Sound. Bright red to brownish-orange; variable in shape, thin layer encrusting shells and rocks when young, maturing to upright irregular fanlike lobes. Pores inconspicuous. Tolerant of pollution and reduced salinity.
Habitat: Bays, estuaries; hard surfaces; lower intertidal to subtidal

COELENTERATES - Primitive animals; radial symmetry; nematocysts; soft bodies either attached polyp or free-swimming medusa.

Moon Jellyfish (*Aurelia aurita*)
Diameter: up to 10 in.
Venom essentially harmless to humans. Body saucer-shaped, 8-lobed, whitish, translucent, fringed with short tentacles. Four white or pinkish horseshoe-shaped gonads in center of disk. Feeds on plankton.
Habitat: Ocean, bays, estuaries; free-floating

Lion's Mane or Red Jelly (*Cyanea capillata*)
Diameter: up to 16 in.
Sting causes itchy, burning rash. Body bell-shaped, smooth, yellowish-orange to reddish-brown; 8 main lobes divided into marginal lobes. Underside has numerous mouth-arms and long, yellowish tentacles in 8 clusters. Sometimes float in large swarms.
Habitat: Ocean, bays, estuaries; free-floating

Frilled Anemone (*Metridium senile*)
Height: up to 4 in.
Largest and most common sea anemone on Long Island. Column smooth, attached at base, orange or yellowish-brown, rarely white or mottled. Abundant fine, white tentacles, not ringed, around mouth cavity; retract when disturbed. When threatened, animal discharges long white threads of stinging cells.
Habitat: Bays, estuaries; on rocks, pilings; intertidal to 500'

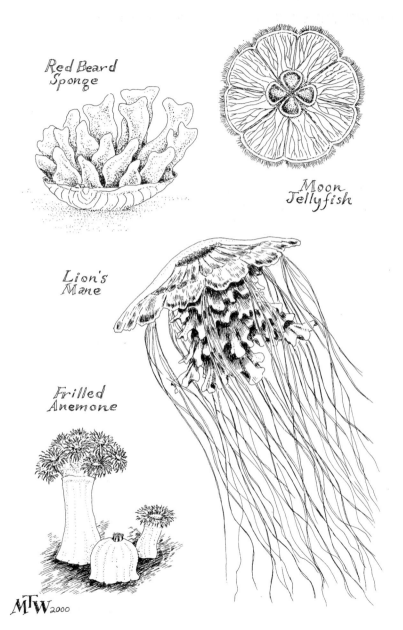

Red Beard
Sponge

Moon
Jellyfish

Lion's
Mane

Frilled
Anemone

MTW 2000

85

CTENOPHORES - Resemble jellyfish but lack nematocysts; biradial symmetry; saclike bodies; 8 rows of ciliary plates; lobed tentacles.

Leidy's Comb Jelly (*Mnemiopsis leidyi*)
Length: up to 4 in.
Luminescent ciliary plates flash bright green when animal disturbed. Lobes longer than oval, translucent body. Propelled by rhythmic beating of cilia. Usually occur in swarms. Predator, zooplankton, occasionally cannibalistic.
Habitat: Shallow water of bays, estuaries, ocean; free-floating

BRYOZOANS - Sessile colonial animals; variable form unique to each species; calcareous exoskeletons.

Spiral-tufted Bryozoan (*Bugula turrita*)
Height: up to 3 in. (rarely to 1 ft.)
Often mistaken for seaweed. Erect, branched, bushy, yellow-orange to tan; light calcification. Zooecia in 2 parallel rows; spiny upper ends; pincers near front margin. Feeds on plankton.
Habitat: Bays, estuaries; on substrate; lower intertidal to subtidal

ANNELIDS - Bristled, segmented worms; terrestrial or aquatic.

Clam Worm or Sandworm (*Nereis virens*)
Length: up to 1 ft.
Popular fishing bait. Body iridescent bluish or greenish-brown with red or gold spots; paler below; 200 segments with parapods; head has 8 tentacles, 4 eyes, 2 sicklelike jaws at end of proboscis. Lives in burrows; fast swimmer. Omnivore, preying on invertebrates, algae, carrion, small fish. Nocturnal.
Habitat: Ocean, bays, estuaries; variable substrate; upper intertidal to 500'

Atlantic Tube Worm (*Hydroides uncinata*)
Length: up to 3 in.
Lives within hard, white, calcareous, snakelike tube, solitary or massed. Body translucent greenish or yellowish, visible blood vessels; head plumed, color variable. Operculum stalked, funnel-shaped, toothed. Filter feeder.
Habitat: Ocean, bays, estuaries; attached to substrate; lower intertidal to 50'

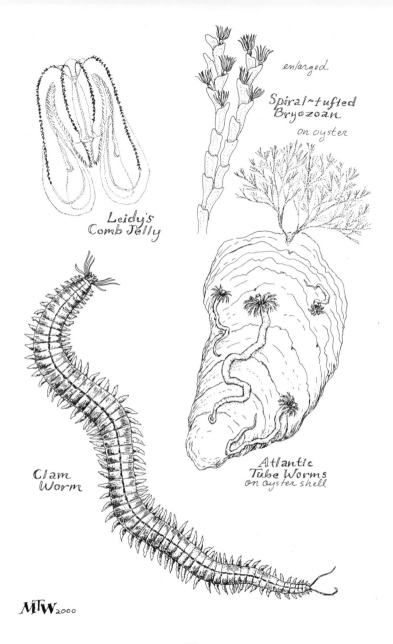

enlarged

Spiral~tufted
Bryozoan

on oyster

Leidy's
Comb Jelly

Clam
Worm

Atlantic
Tube Worms
on oyster shell

MTW₂₀₀₀

87

MOLLUSKS - Soft-bodied, usually with an external shell.

Pelecypods (Bivalves) - Body within two shells connected by a hinge; all filter feeders unless otherwise noted.

False Angel Wing (*Petricola pholadiformis*)
Family: Petricolidae   Size: up to 2 in.
Bores into peat, clay, wood and rock. Shell oblong, thin, chalky-white. Scaly ribs (used for boring) radiate from inflated umbones. External ligament. Interior shiny white with 2 long, pointed cardinal teeth at hinge.
Habitat: Bays, estuaries; hard sediment; intertidal to subtidal

Blood Ark (*Anadara ovalis*)
Family: Arcidae   Size: up to 3 in.
Only red-blooded mollusk on Long Island. Shell ovate, thick, swollen, all white. Exterior with grooved ribs; thick, hairy, dark brown periostracum frequently worn off. Hooked umbones; multitoothed interlocking hinge.
Habitat: Bays, estuaries; sandy bottoms; subtidal

Transverse Ark (*Anadara transversa*)
Family: Arcidae   Size: up to 1.25 in.
More common than *A. ovalis*. Shell trapezoidal, thick, all white, gray-brown periostracum. Exterior ribbed; left valve usually beaded, slightly overlapping right. Umbones straight separated by gap; hinge with ligament.
Habitat: Bays, estuaries; in mud; subtidal

Smooth Astarte or Chestnut Astarte (*Astarte castanea*)
Family: Astartidae   Size: up to 1 in.
Animal has bright red foot; single valves wash up on beach. Shell triangular, thick; exterior brown due to persistent periostracum; smooth with concentric lines; interior white. Hooked umbones; margins crenulate; hinge toothed; small ligament.
Habitat: Ocean; usually in mud; subtidal to 100'

Arctic Wedge Clam (*Mesodesma arctatum*)
Family: Mesodesmatidae   Size: up to 1.5 in.
Single valves usually found. Shell wedge-shaped, thick, whitish to tan. Exterior chalky, compressed, irregular growth lines, yellowish-brown periostracum. Spoon-shaped chondrophore at toothed hinge; umbones toward posterior.
Habitat: Ocean; in sand; subtidal to 300'

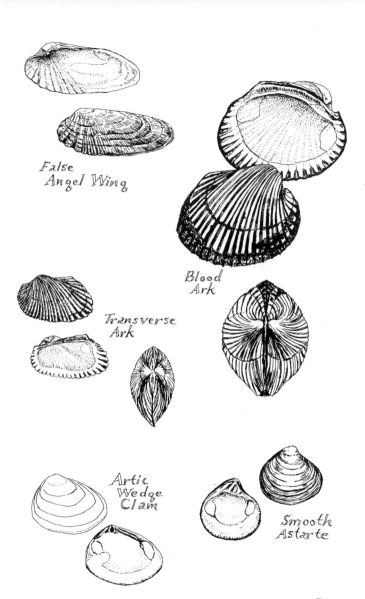

False
Angel Wing

Blood
Ark

Transverse
Ark

Artic
Wedge
Clam

Smooth
Astarte

Atlantic Surf Clam (*Spisula solidissima*)
Family: Mactridae   Size: up to 8 in.
Largest bivalve on Long Island, burrows with tongue-like foot. Shell
broadly triangular. Exterior smooth with fine growth lines, chalky,
white to tan, often with yellow-brown periostracum. Interior glossy
white. Large central umbones; resilium rests on prominent chondrophore
at hinge. Edible.
Habitat: Ocean; in sand; subtidal to 200'

Soft-shelled Clam or Steamer (*Mya arenaria*)
Family: Myidae   Size: up to 5.5 in.
Valuable food source for many animals. Shell elliptical, chalky-white,
moderately thin. Exterior wrinkled; flaky gray-green periostracum.
Flatter left valve with erect chrondrophore which fits into groove on
right valve. Paired shells gape at ends to allow siphons to emerge.
Evident at low tide by jets of water squirting from burrow in sand.
Edible, commercially harvested.
Habitat: Bays, estuaries; in sand, mud; intertidal to 30'

Morton's Egg Cockle (*Laevicardium mortoni*)
Family: Cardiidae   Size: up to 1 in.
Mollusk digs and moves with long, narrow, white foot. Shell thin,
inflated, heart-shaped when viewed from side. Exterior smooth, whitish, frequently with brown zig-zag markings. Interior shiny, bright
yellow, fading to white, with purplish patch on posterior edge. Central
umbones yellowish; hinge with marginal teeth; external ligament.
Habitat: Bays, estuaries; in sandy mud; intertidal to 20'

Jingle Shell (*Anomia simplex*)
Family: Anomiidae   Size: up to 2 in.
Very common, often used in windchimes. Shell thin, glossy, translucent, variable in shape and color (silver, yellow, orange). Upper valve
convex, lower valve flat with hole near hinge to allow calcified
peduncle to attach to hard substrate. Extension from lower valve
supports ligament.
Habitat: Bays, estuaries; on rocks, hard substrate; intertidal to 30'

Balthic Macoma (*Macoma balthica*)
Family: Tellinidae   Size: up to 1.5 in.
Deposit feeder, mainly detritus. Shell variably ovate, flattened, narrower at one end. Exterior whitish with pinkish tinge, grayish
periostracum. Interior pinkish. Central, prominent umbones, usually
worn; hinge with lateral teeth.
Habitat: Brackish water in estuaries; mud, sandy-mud; intertidal to 60'

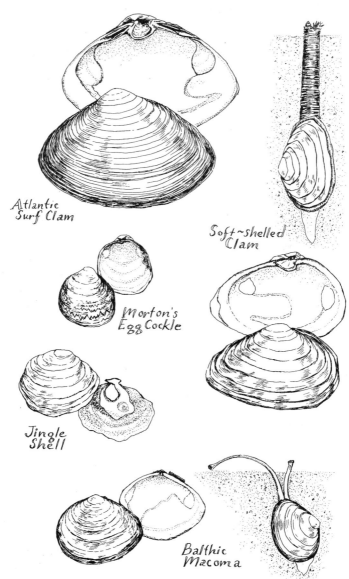

Atlantic
Surf Clam

Soft~shelled
Clam

Morton's
Egg Cockle

Jingle
Shell

Balthic
Macoma

MTW 2000

91

Blue Mussel (*Mytilus edulis*)
Family: Mytilidae   Size: up to 3 in.
Abundant, attaches to substrate with byssal threads. Shell elongate tear-drop, thin. Exterior smooth with prominent growth lines, blue-black, sometimes rayed, with brownish-black periostracum. Interior pearly white, dark purple border, 4-7 small marginal teeth at hinge. Pointed umbones at apex; external ligament. Edible, commercially harvested.
Habitat: Bays, estuaries, ocean; hard substrate; intertidal to subtidal

Ribbed Mussel (*Geukensia demissa*)
Family: Mytilidae   Size: up to 4 in.
Common in salt marsh, partially imbedded in peat. Shell thin, oblong oval. Exterior with radiating ribs, faded gray; shiny yellow-brown periostracum. Interior iridescent bluish-white. Umbones near apex; no hinge teeth.
Habitat: Salt marsh, estuaries, brackish water; in mud; intertidal

Eastern Oyster (*Crassostrea virginica*)
Family: Ostreidae   Size: up to 7 in.
Free-swimming spat, sessile adult cemented to substrate. Shell thick, elongate, variable in shape, flatter upper valve nests into deeper lower half. Exterior rough, layered, dirty-white to gray. Interior smooth, white, purple muscle scar and margin. Umbones elongate, curved. Edible, commercially harvested.
Habitat: Bays, estuaries; hard substrate; intertidal to 40'

Gould's Pandora (*Pandora gouldiana*)
Family: Pandoridae   Size: up to 1.5 in.
Often mistaken for shell fragment. Upper valve flat, lower one convex, extremely compressed, rounded wedge-shape, with squarish ridge on dorsal margin. Exterior smooth with concentric growth lines, chalky-white, often eroded to pearly layer. Interior pearly-white. Tiny umbones at anterior end.
Habitat: Bays, estuaries; in mud; subtidal to 480'

Northern Quahog (*Mercenaria mercenaria*)
Family: Veneridae   Size: up to 5 in.
Also known as Cherrystones and Little Necks. Shell inflated, thick, rounded ovate. Exterior smooth, with many raised concentric growth lines; grayish-white. Interior shiny, white, stained purple, crenulate edge. 3 prominent teeth at each hinge; hooked umbones; heart-shaped lunule. Fragments carved into wampum beads by Native Americans. Edible, commercially harvested.
Habitat: Bays, estuaries; in sand and mud; intertidal to 40'

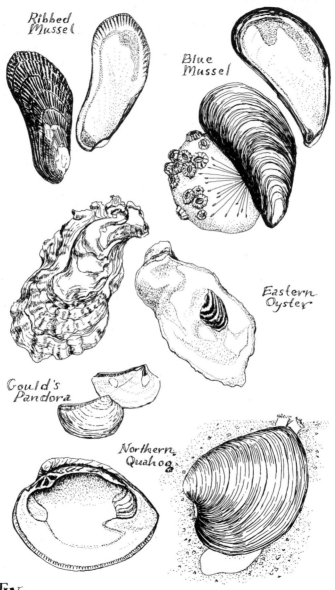

Ribbed
Mussel

Blue
Mussel

Eastern
Oyster

Gould's
Pandora

Northern
Quahog

MTW 2000

93

Common Razor Clam (*Ensis directus*)
Family: Solenidae   Size: up to 8 in.
Also called Atlantic Jacknife Clam. Shell long, rectangular, curved, thin, gaping at ends. Exterior chalky-white; flaky yellowish-green periostracum. Interior bluish-white; 2 cardinal teeth on left valve, 1 on right. Umbones toward anterior end; external ligament. Rapid burrower and swimmer. Edible.
Habitat: Bays, estuaries; in muddy sand; intertidal to subtidal

Atlantic Bay Scallop (*Argopecten irradians*)
Family: Pectinidae   Size: up to 3 in.
Free swimmer; many blue eyes along edge of mantle. Shell fan-shaped, 17-18 elevated ribs, upper valve more convex. Exterior color variable, white to reddish, usually gray. Interior shiny bluish-white. Umbones central, flanked by unequal hinge ears. Lives in eelgrass beds, often attached to substrate with byssus. Edible, commercially harvested, most commonly in Peconic Bay.
Habitat: Shallow bays; eelgrass beds; subtidal to 60'

Deep-sea Scallop (*Placopecten magellanicus*)
Family: Pectinidae   Size: up to 8 in.
Occasionally single valves found on beach. Shell large, thick, nearly round, flattened. Exterior with numerous, thin radiating ribs and concentric growth lines; upper valve tan to reddish-brown; lower valve, pinkish to white. Interior glossy-white. Umbones central, flanked by equal hinge ears. Edible, commercially harvested.
Habitat: Ocean; subtidal to continental shelf

Naval Shipworm (*Teredo navalis*)
Family: Teredinidae   Size: shell up to 1/4 in.
Marine "termite", causing damage by burrowing into untreated wood. Shell tiny, white, with wing-like projections; multi-toothed exterior sculpture. Wormlike animal encased in long, white, variably-sized calcareous tube.
Habitat: Wood pilings, ships

Stout Tagelus (*Tagelus plebeius*)
Family: Solecurtidae   Size: up to 4 in.
Sometimes confused with razor clam. Shell elongate with rounded, gaping ends, parallel margins. Exterior buff-white with brownish-green periostracum; strong concentric growth lines. Interior white; hinge with 2 small cardinal teeth and swollen callus. Umbones inconspicuous, off-center.
Habitat: Bays, estuaries; in muddy sand; intertidal to 25'

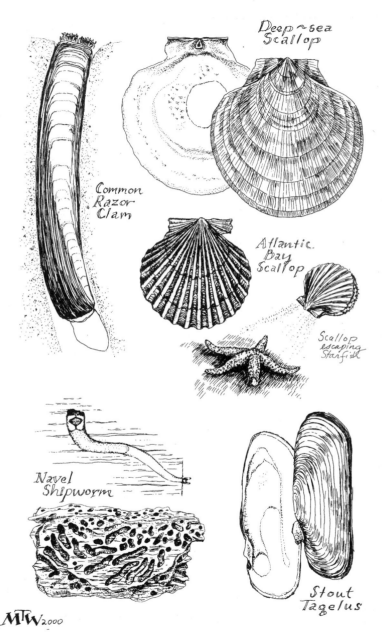

Deep~sea
Scallop

Common
Razor
Clam

Atlantic
Bay
Scallop

Scallop
escaping
Starfish

Navel
Shipworm

Stout
Tagelus

MTW 2000

Gastropods (Univalves) - Single-shelled, usually coiled; most with chitinous operculum. Feed using radula. Largest class of Mollusks.

Atlantic Dogwinkle (*Nucella lapillus*)
Family: Muricidae   Size: up to 2 in.
Common from Montauk to Moriches Inlet. Shell very thick, ovate, fusiform, 5 whorls. Exterior with sculptured ridges; color variable, rarely banded. Aperture oval, thickened outer lip, teeth within, purplish-brown. Siphonal canal short; columella smooth with depression. Elliptic operculum. Carnivorous.
Habitat: Ocean; on rocks; intertidal to subtidal

Atlantic Plate Limpet (*Notoacmaea testudinalis*)
Family: Acmaeidae   Size: up to 1 in. diameter
Primitive gastropod. Shell thin, uncoiled, oval, conical, apex toward anterior end. Exterior smooth, gray, with irregular brown streaks. Interior glossy, bluish-white, central dark brown patch. Herbivorous grazer.
Habitat: Ocean, bays, estuaries; on rocks; intertidal to shallow subtidal

Eastern Melampus (*Melampus bidentatus*)
Family: Melampidae   Size: up to 1/2 in.
Air-breathing snail. Shell ovate-conical, 5-6 whorls, broadest at body whorl; spire short, sometimes eroded. Color olive-brown, shiny, smooth, usually with 3-4 darker bands. Aperture long, narrow, wider at base; outer lip thin, 1-4 lirae within; columella with 2 folds (teeth). Herbivorous, lives in colonies.
Habitat: Salt marsh; on marsh grass; intertidal

Northern Moonsnail (*Euspira heros*)
Family: Naticidae   Size: up to 5 in.
Deposits eggs in circular sand collar. Shell globular, 5 whorls, slightly elevated spire, wide body whorl. Exterior grayish-white to brownish-gray with thin, yellow-brown periostracum. Aperture large, oval, glossy, brown. Umbilicus open. Operculum brown, half-moon shaped. Carnivorous.
Habitat: Ocean; sandy bottoms; subtidal to 1200'

Shark Eye Moonsnail (*Neverita duplicata*)
Family: Naticidae   Size: up to 2.5 in.
Similar to Northern Moonsnail. Shell globular, 4-5 whorls, wider than high. Exterior glossy, smooth, blue-gray to tan, base of shell usually white. Umbilicus deep, almost covered by large, brown callus. Aperture brown. Carnivore.
Habitat: Bays, estuaries; sandy bottoms; intertidal to subtidal

Atlantic
Plate Limpet

Atlantic
Dogwinkle

Eastern
Melampus
to .5 inch

Shark Eye
Moonsnail

Northern
Moonsnail

Operculum

Sand
Collar

Eastern Mudsnail (*Ilyanassa obsoleta*)
Family: Nassariidae   Size: up to 1 in.
Abundant, usually in massive colonies. Shell ovate, pointed spire (often eroded), 5-6 whorls. Exterior blackish-brown, sometimes with white bands; beaded. Aperture oval, shiny purple-brown, thick parietal wall; ridged columella; siphonal notch at base. Operculum oval. Omnivorous scavenger.
Habitat: Bays, estuaries; mud flats; intertidal, subtidal in winter

Three-lined Nassa (*Nassarius trivittatus*)
Family: Nassariidae   Size: up to .75 in.
Also known as New England Nassa. Shell ovate, tall conical spire, 6 shouldered whorls. Exterior yellowish-white, often with brown bands; prominently beaded. Aperture oval, lirae within outer lip; parietal wall white. Operculum serrated. Common colonial scavenger.
Habitat: Ocean, bays, estuaries; in clean sand; intertidal to subtidal

Atlantic Oyster Drill (*Urosalpinx cinerea*)
Family: Muricidae   Size: up to 1.5 in.
Destructive to oyster beds. Shell fusiform, 5-6 whorls. Exterior color variable, usually gray; 10-12 axial ribs crossed by spiral threads. Aperture dark brown, 2-6 whitish teeth within. Siphonal canal straight. Operculum elliptical. Carnivore; drills holes in shells with radula.
Habitat: Bays, estuaries; in rubble; intertidal to subtidal

Thick-lipped Oyster Drill (*Eupleura caudata*)
Family: Muricidae   Size: up to 1 in.
Less common than Atlantic Oyster Drill but equally destructive. Shell almost fusiform, flattened, pointed spire, 5 shouldered whorls. Exterior bluish-gray to brown; vertical ribs crossed by spiral threads; 2 broad varices. Aperture bluish-white, thick with small whitish teeth. Columella curved, white. Siphonal canal deep, almost closed, dark at tip. Operculum oval. Carnivorous.
Habitat: Bays, estuaries; in rubble; intertidal to subtidal

Common Periwinkle (*Littorina littorea*)
Family: Littorinidae   Size: up to 1 in.
Introduced from Europe. Shell rounded, thick, short spire, 6-7 whorls. Exterior grayish-brown to black, dark spiral threads. Aperture brown-black, almost round, thick lip. White columella; no umbilicus. Operculum almost round. Herbivorous grazer. Popular food in Europe.
Habitat: Ocean, bays, estuaries; rocky shore; intertidal to subtidal

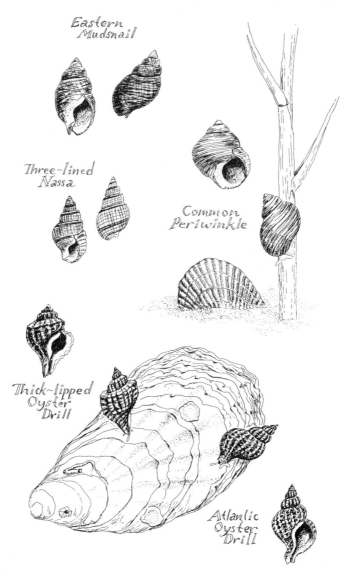

Eastern
Mudsnail

Three-lined
Nassa

Common
Periwinkle

Thick~lipped
Oyster
Drill

Atlantic
Oyster
Drill

M^T_W 2000

Common Atlantic Slipper Shell (*Crepidula fornicata*)
Family: Crepidulidae   Size: up to 2 in.
Most abundant gastropod on Long Island beaches. Shell elongate, variably convex. Exterior white with brown lines and spots. Interior shiny, light brown, dark blotches, covered halfway by wavy-edged white septum. Apex off center, attached to body whorl. No operculum. Herbivorous filter feeder. Live stacked on top of each other attached to hard surfaces; protandric.
Habitat: Bays and estuaries; hard substrate; intertidal to 60'

Eastern White Slipper Shell (*Crepidula plana*)
Family: Crepidulidae   Size: up to 1.5 in.
Attaches to inside of empty gastropods. Shell elongate, thin, flattened, variably convex, white. Exterior crinkled. Interior shiny, septum less than half the length of shell. Apex pointed, centrally located. No operculum. Herbivorous.
Habitat: Bays, estuaries, ocean; inside empty shells; intertidal to subtidal

Channeled Whelk (*Busycotypus canaliculatus*)
Family: Melongenidae   Size: up to 7 in.
Strings of keel-edged egg capsules wash ashore. Shell pear-shaped; 5-6 whorls separated by deep channels; ridged shoulders. Exterior light brown; thick, fuzzy periostracum. Aperture brown. Columella arched; siphonal canal long, straight. Operculum oval, rough. Carnivorous; foot pries open prey.
Habitat: Bays, estuaries; sand; intertidal to 50'

Knobbed Whelk (*Busycon carica*)
Family: Melongenidae   Size: up to 9 in.
Our largest gastropod; similar to Channeled Whelk. Shell pear-shaped; 6 knobbed whorls. Exterior grayish-tan; brown periostracum. Aperture oval, large, color variable. Columella arched; siphonal canal long, straight. Operculum oval, thick. Carnivorous. Ridge-edged egg capsule. Edible (scungili).
Habitat: Bays, estuaries; sand; intertidal to 30'

Chitons - primitive mollusk with 8 calcareous plates held together by a leathery girdle; attaches to rocks; curls when removed from substrate.

Eastern Banded Chiton (*Chaetopleura apiculata*)
Family: Chaetopleuridae   Size: up to 1 in.
Common on eastern Long Island. "Shell" gray to pale brown, flat, oval to oblong. Plates beaded; girdle granulose with short spines. Herbivore.
Habitat: Bays, estuaries; on rocks; intertidal to 90'

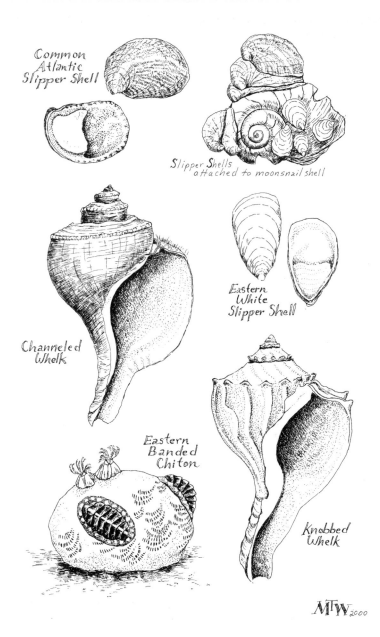

Common
Atlantic
Slipper Shell

Slipper Shells
attached to moonsnail shell

Channeled
Whelk

Eastern
White
Slipper Shell

Eastern
Banded
Chiton

Knobbed
Whelk

MTW 2000

ECHINODERMS - Marine, "spiny skin", radial symmetry, tube feet.

Forbes' Sea Star (*Asterias forbesi*)
Class: Stelleroidea   Size: up to 7 in.
Common "starfish". Body brown, tan, greenish with tones of orange, red or pink; large disk, 5 stout, blunt, cylindrical arms; orange sieve plate; flexible calcareous exoskeleton covered with short, dull spines; 4 rows of tube feet. Carnivore; opens bivalves with arm pressure, then everts stomach into shell.
Habitat: Ocean, bays; rocks, sand; intertidal to 160'

Purple Sea Urchin (*Arbacia punctulata*)
Class: Echinoidea   Size: up to 2 in. wide
Resembles pincushion; spineless exoskeleton washes ashore. Body globular, black comprised of close-fitting calcareous plates (test); covered with movable, blunt, purplish-brown spines up to 1" long. Omnivore/scavenger. **Green Sea Urchin** (*Strongylocentrotus droebachiensis*) (not illustrated) - similar but test larger (3 in.); spines grayish-green, thin, needle-like, 1" long.
Habitat: Ocean, bays; rocks, sand; intertidal to subtidal

ARTHROPODS - "Joint-legged", segmented exoskeletons; growth accompanied by molting.

Saltmarsh Greenhead Fly (*Tabanus nigrovittatus*)
Class: Insecta; Order: Diptera   Body length: up to 1.25 in.
One of the common, bothersome "horseflies". Body dark, stout, broad abdomen; head large, flattened; short antennae; protruding, iridescent green eyes. Aquatic larvae. Similar to mosquitoes: only females are biting "bloodsuckers" for egg production; males drink nectar.
Habitat: Salt marsh, beach

Saltmarsh Mosquito (*Aedes sollicitans*)
Class: Insecta; Order: Diptera   Length: up to .25 in.
Abundant in tidal wetlands. Body brownish, white stripe on top of pointed abdomen; wings clear with dark and light scales; maxillary palp much shorter than white-banded, straw-like, piercing proboscis; male's antennae feathery, female's threadlike. Females draw blood for protein needed in egg development; males drink nectar. Aquatic larvae, called wrigglers, live just below surface of water. Multiple generations per year.
Habitat: Salt marsh

Forbes'
Sea Star

Purple
Sea Urchin

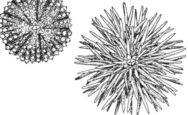

Saltmarsh
Greenhead
Fly

Saltmarsh
Mosquito

MTW 2001

Horseshoe Crab (*Limulus polyphemus*)
Class: Merostomata   Length: up to 24 in. (female), 15 in. (male)
Harmless "living fossil", dating back to over 350 million years ago; related to spiders. Body greenish-brown, broad carapace, horseshoe-shaped; bulbous eyes; triangular abdomen with spiked edges; long, pointed tail (used for flipping over); 5 pairs walking legs surround mouth. Predator, bulldozes in sand for mollusks, worms. Lays eggs on shore late spring.
Habitat: Bays, estuaries; variable substrates; subtidal to 75'

Northern Rock Barnacle (*Balanus balanoides*)
Class: Crustacea   Size: up to 1 in. high
Abundant, usually in crowded colonies. Shell variable, generally conical, whitish (frequently green from algae), sharp-edged on top; comprised of 4 rough, pleated, slightly overlapping calcareous plates cemented to substrate; 4 smaller "trapdoor" plates protect retractable feathery feet. Omnivore, plankton, detrital matter, gathered by feeding appendages. Free-swimming larvae.
Habitat: Ocean; rocky substrate; intertidal to subtidal
Barnacle species are adapted to specific salinities. The **Ivory Barnacle** (*B. eburneus*), with smooth, white shell plates, occurs in estuarine conditions. The similar **Bay Barnacle** (*B. improvisus*) is found in brackish to nearly freshwater. (not illustrated)

Beach Flea (*Orchestia* spp.)
Class: Crustacea; Order: Amphipoda   Length: up to .75 in.
Ubiquitous "sand hopper"; hides under washed up vegetation, springs upward when disturbed. Body shiny, elongated, laterally compressed, brown or buff; long, hairy antennae; 7 pairs of legs, last 3 pairs adapted for leaping; telson curved downward. Harmless voracious scavenger.
Habitat: Intertidal to upper beach

Atlantic Mole Crab (*Emerita talpoida*)
Class: Crustacea   Length: up to 1 in.
Modified for life on wave-swept beaches, constantly burrowing and being uncovered again. Body egg-shaped, grayish-tan to white; tapering abdomen; legs clasped tightly below carapace; long eyestalks; 4 antennae, one feathery pair strains water for food; lacks pincers. Harmless scavenger, organic debris.
Habitat: Sandy ocean beach; intertidal

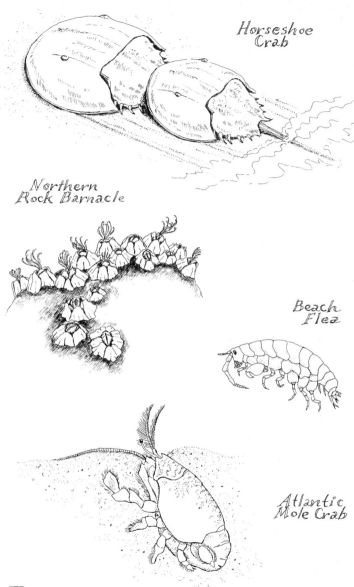

Horseshoe
Crab

Northern
Rock Barnacle

Beach
Flea

Atlantic
Mole Crab

MTW₂₀₀₀

True Crabs - 5 pairs of well-developed legs, 1st for grasping (claws), remainder for locomotion; usually walk sideways; eyes stalked; hard, toothed carapace; large cephalothorax, small abdomen; most scavengers, but will eat live prey.

Atlantic Rock Crab (*Cancer irroratus*)

Class: Crustacea   Width: up to 5 in.
Edible. Carapace oval, yellow, heavily freckled with reddish or purplish spots; 9 wide, rounded marginal teeth on each side. Pincers stubby, stout, black-tipped, point downwards. Legs short, hairy-edged. Pests to lobstermen.
Habitat: Sound, ocean; over rocks, gravel, sand; intertidal to 2,600'

Blue Crab (*Callinectes sapidus*)

Class: Crustacea   Width: up to 9 in.
Important commercial species; edible. Carapace bluish-green, smooth; each side with sharp, red-tipped marginal teeth, the 9th a strong spine. Sharp, powerful claws, blue (male), red-tipped (female). Legs blue, last pair paddle-shaped for swimming. Male's abdominal apron narrow, female's wide triangle. Easily provoked; pinch painful.
Habitat: Estuaries, salt marsh, ocean; eelgrass, mud bottoms; subtidal to 120'

Calico Crab or Lady Crab (*Ovalipes ocellatus*)

Class: Crustacea   Width: up to 3 in.
Burrows in sand, waiting in ambush for prey. Carapace fan-shaped, 5 marginal teeth each side, yellowish-gray covered with rings of purplish-red spots. Pincers strong, large, with purplish blotches. Legs spotted, last pair paddle-like for swimming. Aggressive.
Habitat: Bays, sounds, ocean; sand, mud, rock bottoms; subtidal to 150'

Fiddler Crab (*Uca* spp.)

Class: Crustacea   Width: up to 1 in.
Colonial; lives in 2' burrow. Male distinguished by one greatly enlarged claw for mating display and fighting; female with 2 small claws. Carapace trapezoidal. Long, thin eyestalks. Legs hairy, adapted for walking. Feed on bacteria, algae from sediment, occasionally cannibalistic. **Sand Fiddler** (*U. pugilator*) has smooth claw; carapace purplish with black markings. **Mud Fiddler** (*U. pugnax*) has tubercles on claw; carapace olive to black, bluish in front.
Habitat: Salt marsh, intertidal mudflats; 2 species overlap in territory, *U. pugnax* prefers mud, *U. pugilator* favors sand.

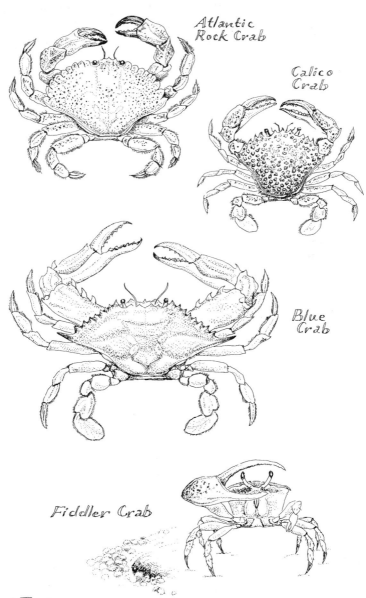

Atlantic
Rock Crab

Calico
Crab

Blue
Crab

Fiddler Crab

MTW2000

107

Green Crab (*Carcinus maenas*)
Class: Crustacea   Width: up to 3 in.
Introduced from Europe. Carapace fan-shaped, green with black mottling above; males and immature yellow or green below, females orange-red; 5 sharp marginal and 3 frontal teeth each side. Pincers large, equal; 5th pair legs flattened, not paddle-shaped, although crab swims. Tolerant to low salinity.
Habitat: Salt marsh, tidal pools; rocks, jetties; intertidal to subtidal

Mud Crab (*Eurypanopeus depressus*)
Class: Crustacea   Width: up to .75 in.
Predator on young clams and oysters, often hiding in empty shells. Carapace flattened, wide, rounded in front, olive-brown, lighter underneath; 5 marginal teeth, first 2 fused; area between eyes almost straight with central notch. Pincers powerful, unequal, black-tipped, fingers of minor claw spoon-shaped.
Habitat: Bays, estuaries; under rocks, debris; intertidal

Common Spider Crab (*Libinia emarginata*)
Class: Crustacea   Width: up to 4 in.
Leg span can reach 1 ft. Carapace round, strongly tubercled, hairy, brown or dull-yellow; row of spines down back. Pincers white-tipped, long, equal; legs round, hairy. Usually covered with algae and debris, often placed there by crab for camouflage. Harmless, slow-moving.
Habitat: Ocean, bays, estuaries; various bottoms; intertidal to 400'

Hermit Crabs - Live in empty snail shells to protect long, soft, cylindrical abdomens. Rasps on telson and uropods hold animals in shell.

Long Clawed Hermit Crab (*Pagurus longicarpus*)
Class: Crustacea   Length: up to .5 in.
Small, extremely abundant in shallows. Carapace widest at rear, oblong, gray, green or buff. Pincers unequal; major claw narrow, cylindrical, often with dark median stripe; palm nearly hairless and smooth. 2 pairs strong walking legs, 2 other pairs underdeveloped. Occupied shell frequently fuzzy in appearance due to presence of **Snail Fur** (*Hydractinia echinata*) a colonial, reddish-brown hydroid. Polymorphic polyps with short, erect spines, emerge from a brownish crust.
Habitat: Bays, estuaries, ocean; variable bottoms; intertidal to 150'

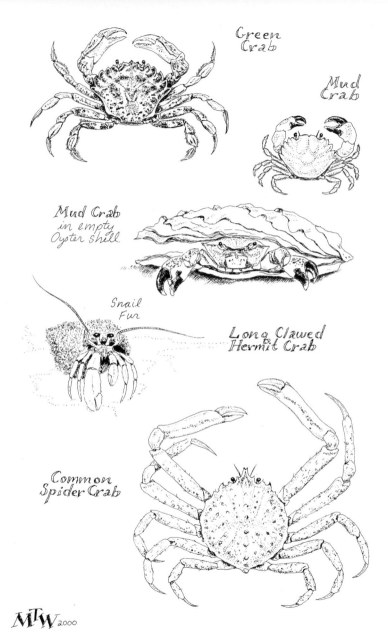

Green
Crab

Mud
Crab

Mud Crab
in empty
Oyster shell

Snail
Fur

Long Clawed
Hermit Crab

Common
Spider Crab

MTW 2000

Northern Lobster (*Homarus americanus*)
Class: Crustacea   Length: up to 36 in.
Highly prized delicacy. Thorax and segmented abdomen cylindrical, reddish-brown to greenish-brown, paler underneath. 3 pairs of red-tipped pincers, 1st pair large, powerful, unequal; rear 2 walking legs smaller. Head pointed, two sets of red antennae, 1 pair longer than body; round black eyes on stalks. Telson red, flattened, fan-shaped. Omnivorous, sometimes cannibalistic. Molts annually. Fished commercially using cagelike traps.
Habitat: Ocean, Long Island Sound; rocky, sandy bottoms; subtidal to continental shelf

Common Mantis Shrimp (*Squilla empusa*)
Class: Crustacea   Length: up to 10 in.
Ornate, secretive, digs elaborate burrows. Body color variable, usually greenish-gray, darker segment margins; somewhat flattened; short carapace, long segmented abdomen. 5 pairs of walking legs, 2nd pair large strong multitoothed mantislike predatory claws, last 3 pairs small, weak. 2 sets of antennae, 2nd pair long. Flattened telson. Carnivorous. Edible.
Habitat: Ocean, Long Island Sound, bays; mud, sand; lower intertidal to 500'

Sand Shrimp (*Crangon septemspinosa*)
Class: Crustacea   Length: up to 2 in.
Camouflaged, translucent, buff color and mottled pattern resembles sand. Body flattened; dorsal spine. 1st set of legs stout with poorly developed claw, remaining pairs slender. Short beak; telson blackish, flat, fanlike for swimming. Burrowing scavenger. Tolerates wide range of salinity.
Habitat: Bays, estuaries, tidepools; sand, eelgrass; subtidal to 300'

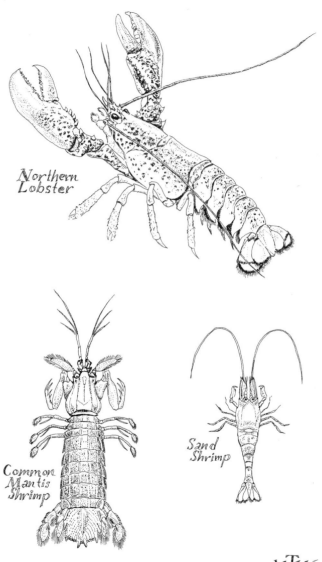

Northern Lobster

Common Mantis Shrimp

Sand Shrimp

MTW 2000

111

VERTEBRATES - Animals that have a backbone.

FISH - Aquatic, cold-blooded; with fins, scales(usually); breathe by gills.

Sand Shark (*Carcharhinus plumbeus*)
Length: up to 8 ft.
Sluggish, bottom-dwelling, cartilaginous. Body dark gray above, whitish below; no markings; ridge between 2 dorsal fins; first fin large, begins above middle of pectoral fin; snout fairly short, rounded; teeth weakly serrated. Carnivore, scavenger, mainly fish, mollusks. Not a threat to people.
Habitat: Shallow bays, estuaries, coastal waters; 10-30' deep

Clearnose Skate (*Raja eglanteria*)
Length: up to 36 in.
Common in summer, bottom-dwelling, cartilaginous. Body flattened, wider than long, light to dark brown above, dark spots, whitish below; single row of spines along back; snout pointed; long, spiny tail. Swims moving winglike pectoral fins. Egg case black "mermaid's purse". Carnivore, fish, crustaceans; hunts at night. Rests half-buried in sand during day.
Habitat: Ocean, estuaries, sound; sandy bottoms; subtidal to 200'

Black Sea Bass (*Centropristis striata*)
Length: up to 2 ft. (usually 15 in. or less)
Elongate, bluish-black (female brownish), mottled with stripes and white blotches; large head; long, spiny, continuous black and white banded dorsal fin; tail fin 3-lobed. Carnivore, crustaceans, fish.
Habitat: Harbors, jetties; rocky bottoms; shallow water to continental shelf

Striped Bass (*Morone saxatilis*)
Length: up to 6 ft. (usually 2 ft.)
Popular sport fish. Streamlined, olive or bluish above, silvery sides with black stripes, white belly; dorsal fins triangular, separate; mouth large; tail fin notched. Carnivore, fish, crustaceans. Anadromous, spawning in freshwater.
Habitat: Ocean, Long Island Sound; deeper water in winter

Blackfish or Tautog (*Tautoga onitis*)
Length: up to 3 ft. (usually 10-15 in.)
Chunky, brownish or dull black, sides mottled lighter brown (female blackish and buff); head rounded, lips thick; long dorsal fin notched and curving at rear; tail fin squared off, slightly rounded. Carnivore, mostly mollusks.
Habitat: Ocean, estuaries, sound; near docks, mussel beds, rocky shores

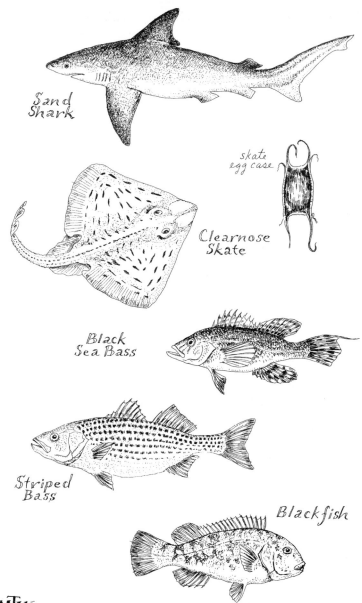

Sand
Shark

skate
egg case

Clearnose
Skate

Black
Sea Bass

Striped
Bass

Blackfish

MTW 2000

113

Bluefish (*Pomatomus saltatrix*)
Length: up to 45 in.
"Piranha of the sea". Elongate, blue-green above, silvery below; black blotch at base of pectoral fin; spines on 1st dorsal fin; large head, wide mouth, razor sharp teeth; tail deeply forked. Voracious schooling predator. Caution: feeding frenzies common and should be avoided by swimmers.
Habitat: Ocean, bays, estuaries; deeper waters in winter

Flatfishes - Bottom-dwelling; when young, eye migrates to join other on one side; continuous dorsal and anal fins fringe flat, elliptical body.
Winter Flounder (*Pleuronectes americanus*)
Length: up to 20 in.
Color variable (with substrate) pale brown, olive, or reddish, often mottled, underside white; bulbous eyes on right side of head, small mouth; lateral line straight; tail fin rounded. Carnivore, crustaceans, fish, mollusks, worms.
Habitat: Coastal waters; muddy and sandy bottoms; deeper waters in summer

Fluke or Summer Flounder (*Paralichthys dentatus*)
Length: up to 30 in.
Body brown with large dark spots, white underneath; large eyes on left side of head; lateral line arched; tail fin wedge-shaped. Buries in sand, darting up quickly in pursuit of food. Carnivore, mostly fish, crustaceans, squid.
Habitat: Bays, harbors, estuaries, ocean; sandy or muddy bottoms

Mummichog (*Fundulus heteroclitus*)
Length: up to 4 in.
"Minnow" bait fish. Stocky, bluish green, silver and dark bars on side; rounded back, spineless dorsal fin set far back; blunt head; large, rounded tail fin. Carnivore, insect larvae, small invertebrates at surface. Sometimes found in brackish waters with **Banded Killifish** (*F. diaphanus*)(not illustrated), a freshwater minnow, similar but narrower, more conspicuous bars on sides.
Habitat: Estuaries, tidal creeks, salt marsh; mud bottoms; shallow water

Northern Pipefish (*Syngnathus fuscus*)
Length: up to 8 in.
Related to seahorse. Flexible body long, cylindrical, dark yellowish-green, mottled, covered in rings of bony scales; rectangular dorsal fin; narrow head, long tubular snout, red eyes; fan-like tail. Carnivore, small invertebrates, tiny fish, plankton. Male carries eggs in brood pouch.
Habitat: Bays, estuaries, inlets; eelgrass; shallow waters

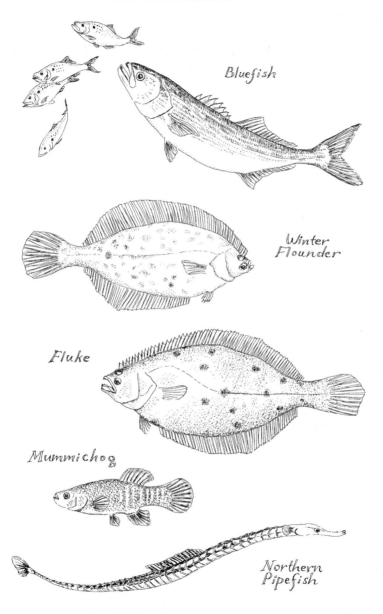

Bluefish

Winter
Flounder

Fluke

Mummichog

Northern
Pipefish

MTW

115

Porgy or Scup (*Stenotomus chrysops*)
Length: up to 16 in.
Common bottom feeder. Body oval, brownish above, dull silvery sides with irregular dark bars, pale blue flecks; blue stripe along base of spiny dorsal fin; head profile steep, front teeth incisorlike; tail fin deeply forked; large scales. Carnivore, worms, crustaceans.
Habitat: Ocean, estuaries; sandy bottoms; piers, jetties

Lined Seahorse (*Hippocampus erectus*)
Length: up to 5 in.
Pipefish family; swims upright. Body long, armored with lines and ridges, color varies with surroundings (brown, gray, yellow or reddish with pale blotches); head and tubular snout angle down; dorsal fin fan-shaped; tail prehensile, curls around vegetation. Carnivore, small crustaceans, fish larvae. Male carries eggs in brood pouch.
Habitat: Ocean, estuaries; eelgrass; shallow waters

Northern Searobin (*Prionotus carolinus*)
Length: up to 15 in.
Distinctive appearance. Body mottled, grayish or reddish above, pale below; head large, broad, flat, bony plate and spines, eyes on brow ridge; triangular dorsal fin with black spot; winglike pectoral fins with 3 free rays used for "walking" and probing for food on bottom. Carnivore, bivalves, crustaceans.
Habitat: Bays, estuaries, sound; shallow waters

Atlantic Silverside (*Menidia menidia*)
Length: up to 5 in.
Schooling fish. Body elongated, greenish above, whitish below, prominent silver stripe with faint black line on sides; small head, large eyes; widely separate dorsal fins, first with spiny rays; tail fin forked. Carnivore, insect larvae, tiny invertebrates. Important food for terns, larger fish.
Habitat: Estuaries, bays, inlets; off sandy beaches

Weakfish (*Cynoscion regalis*)
Length: up to 3 ft. (usually 18 in. or less)
Bottom dweller; makes loud croaking noise by vibrating muscles around air bladder. Body spindle-shaped, dark olive above, silvery below, iridescent sides with small dark spots; pointed snout, lower jaw projects; separate soft dorsal fins; tail fin notched. Carnivore, small fish. Fishing hook easily tears "weak" mouth.
Habitat: Estuaries, sound; sandy bottoms; shallow waters, deeper in winter.

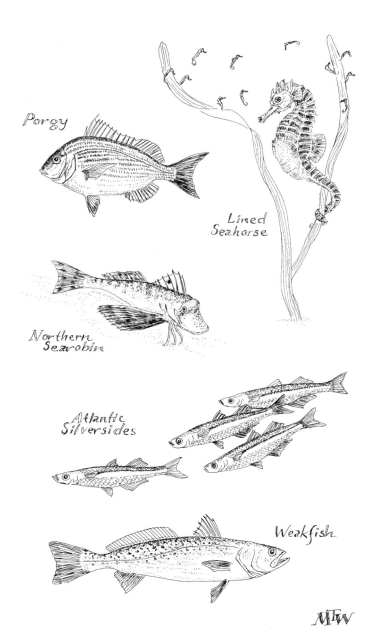

Porgy

Lined
Seahorse

Northern
Searobin

Atlantic
Silversides

Weakfish

MFW

REPTILES - Cold-blooded, with dry, scaly skin, sharp claws (if legged); eggs in leathery covering.
TURTLES - Body covered by bony shell; mouth toothless with horny beak; eggs buried in sand.

Diamondback Terrapin (*Malaclemys terrapin*)
Shell length: up to 8 in. (female), 5 in. (male)   Migratory
Long Island's only reptile restricted to coastal waters. Carapace brown to black, oval, keeled; scutes with concentric ridges; plastron unhinged, yellowish with dark blotches. Body light gray with black spots. Omnivore, mainly fish, invertebrates, occasionally plants. Once popular as food, now protected.
Habitat: Estuaries, salt marsh, lagoons, coastal waters.

Sea Turtles - Marine; legs modified as flippers; omnivorous, invertebrates, especially jellyfish, sometimes vegetation.

Atlantic Green Turtle (*Chelonia mydas*)
Shell length: up to 4 ft. Weight: up to 500 lbs.   Migratory
Largest hard-shelled sea turtle. Carapace olive-brown to black, irregularly patterned, oval, slightly elongate, smooth-edged, 4 costal plates on each side of shell, nuchal separate from costal; plastron yellowish-white. Name derived from body fat color. Threatened.
Habitat: Ocean, inlets, bays, estuaries

Atlantic Ridley Turtle or Kemp's Ridley (*Lepidochelys kempi*)
Shell length: up to 28 in. Weight: up 100 lbs.   Migratory
Smallest and most endangered of our sea turtles. Carapace olive-brown, heart-shaped, 5 costals each side of shell, nuchal touches costal; plastron yellow. Long Island critical habitat for juveniles.
Habitat: Ocean, bays, estuaries, lagoons

Loggerhead Turtle (*Caretta caretta*)
Shell length: up to 3.5 ft. Weight: up to 300 lbs.   Migratory
Most common of the sea turtles; still nests on Atlantic coast. Carapace reddish-brown, flat, elongated, 5 or more costals each side, nuchal touches costal; plastron yellow. Head large, block-like, topped by dark-brown scales; beak pointed. Threatened. **Leatherback** (*Dermochelys coriacea*) - largest sea turtle, up to 8 ft., 2000 lbs.; another common visitor to Long Island's waters (not illustrated).
Habitat: Ocean, estuaries

Diamondback
Terrapin

Atlantic
Ridley
Turtle

Atlantic Green
Turtle

Leatherback
Turtle

MTW 2001

BIRDS - Warm-blooded, with feathers, two feet, wings, toothless beak.

Red-winged Blackbird (*Agelaius phoeniceus*)
Length: up to 9.5 in.   Migratory
Most common wetlands perching bird. Males black; distinctive red shoulder patch with yellow border on bottom; females and young brownish and heavily streaked. Song gurgling "kong-a-ree". Males arrive in late February to claim nesting sites. Omnivore, mainly seeds, some insects.
Habitat: Salt marshes, shorelines

Bufflehead (*Bucephala albeola*)
Length: up to 15 in.   Migratory - rare in summer
Small, plump diving duck; large, round head, short bill. Males mainly white, greenish-purplish head, black back, distinctive white wedge-shaped patch on back of head; females gray with whitish elongated patch on cheek. Male's call whistle, female's soft quack. Fast flyer. Omnivore.
Habitat: Bays, estuaries, ocean

American Coot (*Fulica americana*)
Length: up to 15 in.   Migratory - abundant in winter
Ducklike aquatic bird. Grayish black, prominent white bill, white patch under tail; greenish legs, flanged feet, lobed toes. Call "kuk-kuk-kuk-kuk". Expert diver. Must patter on water to take off. Omnivore, mainly plants.
Habitat: Salt marshes, bays, inlets

Double-crested Cormorant (*Phalacrocorax auritus*)
Length: up to 33 in.   Migratory
Large, dark aquatic bird often perched with wings outspread, drying plumage which is not water-repellant. Glossy, greenish-black body, long slender neck, orange throat patch; slim, hooked bill; 2 curly black crests on head; black feet; immature brownish, whitish throat and breast. Usually silent. Flocks fly in "V". Carnivore, diving from surface for fish.
Habitat: Coasts, bays

Fish Crow (*Corvus ossifragus*)
Length: up to 17 in.   Resident
Large perching bird. Black overall, thin bill; stout feet; long, flattish tail. Slightly smaller and plumage glossier than American Crow. Call nasal "kwok" and 2-noted "anh-anh". Omnivore, mainly insects, bird eggs, fish.
Habitat: Coastlines, salt marshes

Bufflehead

Red~winged
Blackbird

Double~crested
Cormorant

American
Coot

Fish
Crow

MTW 2001

Common Goldeneye (*Bucephala clangula*)
Length: up to 20 in.   Migratory - winter resident
Diving duck, with puffy head, bright yellow eyes. Male head glossy
dark green, large white spot near bill; black and white back, all white
below; female head dark brown, white neck ring, body grayish. Male's
call piercing "spear-spear"; female's quack. Wings whistle in flight.
Omnivore, mollusks, plants.
Habitat: Bays, inlets, tidal rivers

Great Black-backed Gull (*Larus marinus*)
Length: up to 30 in.   Resident
Largest gull. White head and body, black wings; heavy yellow bill with
red dot; pink legs and feet. Immature mottled brown; dark bill; black
tip on tail; takes 4 years to reach adult plumage. Call loud, gutteral
"keeow". Omnivore, scavenger, fish, eggs, refuse.
Habitat: Coastal waters, shores

Herring Gull (*Larus argentatus*)
Length: up to 25 in.   Resident
Most common gull. White head and body, silvery gray wings and back;
wingtips black with white spots; yellow bill with red dot; pink legs and
feet. Immature and winter adult brown flecked. Call varied, loud,
raucous. Omnivore, scavenger, bivalves, fish, garbage; drops shellfish
on rocks to open.
Habitat: Coastlands, ocean, tidal rivers, estuaries, suburban areas

Laughing Gull (*Larus atricilla*)
Length: up to 17 in.   Migratory
Common small gull. Dark gray back and wings; black wingtips; white
neck and underparts; black head in summer (grayish in winter); white
eye ring; red bill; dark orange legs and feet. Immature gray-brown,
black bill. Call high, laughing "ha-ha-ha". Omnivore, fish, insects,
crustaceans.
Habitat: Beach, salt marsh, tidal rivers, ocean, bays, estuaries

Glossy Ibis (*Plegadis falcinellus*)
Length: up to 23 in.   Migratory
Colonial, gregarious wading bird, appearing black at a distance. Dark
brown body; iridescent green wings; purplish-brown head, neck and
underparts; long legs; large, slender, downcurved bill. Call low, nasal
"ka-onk". Carnivore, probes for invertebrates and snakes. Outstretched
neck in flight.
Habitat: Marsh, shallow water

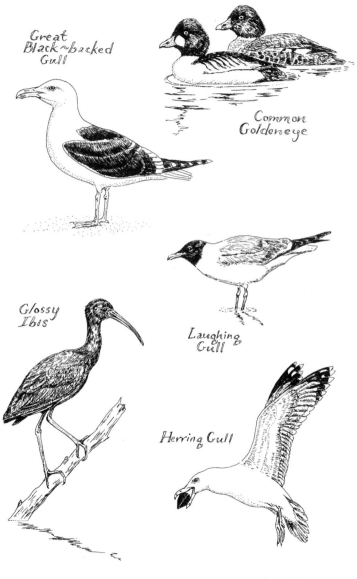

Great
Black~backed
Gull

Common
Goldeneye

Glossy
Ibis

Laughing
Gull

Herring Gull

MTW 2001

Heron family - Wading birds with daggerlike bills, legs trail in flight.

Great Egret (*Ardea alba=Casmerodius albus*)
Length: up to 40 in.   Migratory
Large heron; pure white body, yellow bill, black legs and feet. Breeding plumes extend beyond tail. Call deep hoarse croak. Hunts in shallows by stalking with neck outstretched. Carnivore, mainly fish, frogs, crustaceans.
Habitat: Salt marsh, shorelines, bays, estuaries

Snowy Egret (*Egretta thula*)
Length: up to 27 in.   Migratory
Smallish heron; white body, black bill, black legs, golden yellow feet. Plumes on back long and lacy during breeding. Call low squawk. Hunts by shuffling feet and sprinting in shallows. Carnivore, mainly fish. Almost hunted to extinction for plumes, now protected.
Habitat: Salt marsh, intertidal flats

Black-crowned Night Heron (*Nycticorax nycticorax*)
Length: up to 28 in.   Migratory
Medium-sized, thickset, short-necked; black back and crest, short black bill; white below, gray wings, pale legs. White plumes from back of head during breeding. Call distinctive barking "guawk". Hunts at night moving briskly with lowered head. Carnivore, mainly fish, crustaceans, frogs, mollusks.
Habitat: Salt marsh, intertidal flats

Great Blue Heron (*Ardea herodias*)
Length: up to 52 in.   Migratory - occasional resident
Largest heron; very common; enormous wingspan. Bluish-gray body, whitish head and neck, black stripe over eyes, yellowish bill, black legs. Black plumes from crown during breeding. Call harsh "grahnk" when alarmed. Hunts in shallows, walking slowly or standing with head back on shoulders. Carnivore, mainly fish, frogs.
Habitat: Salt marsh, shorelines, bays, estuaries

Green Heron (*Butorides virescens=Butorides striatus*)
Length: up to 22 in.   Migratory
Small, dark green-gray body; chestnut head and chest, black crown, long dark green and yellow bill, short yellow or orange legs. Call sharp, piercing "skew" when alarmed. Hunts sitting with neck drawn in, then stabs prey after a few cautious steps. Carnivore, mainly fish, salamanders, insects, crayfish.
Habitat: Salt marsh, bays, estuaries

Black-crowned
Night Heron

Great
Blue Heron

Great
Egret

Snowy
Egret

Green
Heron

MTW 97

125

Belted Kingfisher (*Ceryle alcyon*)
Length: up to 13 in. Resident
Pigeon-sized diving bird. Large, crested, blue-gray head; white collar and belly, blue-gray back and upper chest band (female with lower rusty band); large dagger-like bill. Call a loud rattle. Perches along waterways, then hovers and dives for food. Carnivore, mainly fish, crabs, lizards, mice, insects.
Habitat: Coasts, estuaries, salt marsh

Common Loon (*Gavia immer*)
Length: up to 32 in. Migratory - rarely in summer
Very large duck-like bird. Black back with white pattern and spots; black head and stout pointed bill; black neck with prominent neck ring; red eyes; white underside. In winter, upper parts dark gray. Call loud yodel, often at night. Rides low in water when swimming; excellent diver. Carnivore, fish, frogs.
Habitat: Ocean, bays in winter

Hooded Merganser (*Lophodytes cucullatus*)
Length: up to 19 in. Migratory - rarely in summer
Small diving duck. Male dark, blackish body, rusty sides; black-bordered white crest; slender tooth-edged bill; female gray-brown, darker head and crest. Call low grunt. Springs from water at takeoff. Carnivore, mainly fish, frogs, aquatic insects.
Habitat: Tidal rivers, bays, estuaries, salt marsh

Red-breasted Merganser (*Mergus serrator*)
Length: up to 23 in. Migratory - rarely in summer
Diving duck. Male greenish-black head, shaggy double crest, long, narrow red bill; wide white collar, black back, gray sides, white belly, rusty, streaked breast; female rusty head, gray body. Rarely vocal. Carnivore, mainly fish, crustaceans, captured in swift underwater pursuit.
Habitat: Coastal waters, ocean, bays, estuaries

Northern Mockingbird (*Mimus polyglottos*)
Length: up to 10 in. Resident and Migratory
Common, aggressive, territorial mimic. Gray above, paler gray below; 2 white bars and large patch on wings; small, thin bill; long, blackish tail with white outer feathers. Call loud "chack", song imitates other birds; often sings at night. Omnivore, invertebrates and fruit.
Habitat: Salt marsh, maritime forest, dunes, swale

Belted
Kingfisher

Common
Loon

winter

Hooded
Merganser

♂

Red-breasted
Merganser

♀

♂

Northern
Mockingbird

MTW 2001

127

Osprey or Fish Hawk (*Pandion haliaetus*)
Length: up to 24 in.   Migratory
Predator, 70" wingspan, often seen nesting on elevated human-made platforms. Brownish-black above, white below (female's chest has dark markings), black line through eye and side of white head; conspicuous crook in wings when flying. Call loud chirping "kyew-kyew." Hunts hovering over water, plunging feet-first. Carnivore, almost exclusively fish.
Habitat: Salt marsh, estuaries, coastland

American Oystercatcher (*Haematopus palliatus*)
Length: up to 19 in.   Migratory
Recently extended its range to Long Island. Dark brown back, white belly and wing stripe, black head; conspicuous long, straight, bright red bill; pinkish legs. Call loud, harsh "kleep". Carnivore, mainly bivalves; hammers and stabs shells with bill to open.
Habitat: Beach, mud flats, intertidal zone

Black-bellied Plover (*Pluvialis squatarola*)
Length: up to 12 in.   Migratory
Winter visitor, summers in arctic. Plumage grayish, darker above than below in winter. Distinctive white rump, black axillaries (wing pits), wide white wing stripe. Dark gray to black legs and beak. Call a sorrowful three-note whistle. Omnivore, mainly invertebrates.
Habitat: Salt marsh, tidal flats

Piping Plover (*Charadrius melodus*)
Length: up to 7 in.   Migratory
Small shorebird; threatened species, protected nesting areas on Long Island beaches. Pale whitish below, very pale grayish above; black neck band; yellow bill with black tip; orange-yellow legs. Immature and in winter no neck band, all black bill. Call whistled "peep-lo". Carnivore, tiny crustaceans, insects.
Habitat: Barrier beaches, coastlands

Semipalmated Plover (*Charadrius semipalmatus*)
Length: up to 7 in.   Migratory
Common shorebird. Dark brown above, white below; black neck band; yellow bill with black tip, yellowish legs. Immature and winter brown neck band, black bill. Call plaintive "chu-weet". Runs along shore for few paces, then stops and raises head. Carnivore, tiny crustaceans, insects.
Habitat: Beach, tidal flats, salt marsh

Osprey

American
Oystercatcher

Black~bellied
Plover

Semipalmated
Plover

Piping Plover

MTW 2001

Clapper Rail (*Rallus longirostris*)
Length: up to 14 in.   Migratory
Elusive shorebird, more often heard than seen. Grayish-brown body, sides lightly barred; chest and head lighter; white patch under short tail; toes long; bill long, yellowish, slightly downcurved. Call clattering "kek-kek-kek", usually at dawn or dusk. Carnivore, mainly fiddler crabs, snails.
Virginia Rail (*Rallus limicola*) (not illustrated) - similar but smaller (10 in.); reddish bill, rusty underparts; common but elusive.
Habitat: Salt marsh

American Redstart (*Setophaga ruticilla*)
Length: up to 5 in.   Migratory
Warbler. Black with orange patches on wings and sides, white belly (female olive-gray with yellow patches). Flits about with tail fanned. Song series of high tones, ending with higher or lower note. Omnivore, mostly insects, especially flies.
Habitat: Maritime forests

Sanderling (*Calidris alba*)
Length: up to 8 in.   Migratory, often overwinters
Well-known sandpiper; usually in groups. Rusty, mottled head; chest, back, belly white; in winter, pale gray above, white below; legs black; bill short, black. Call sharp "plic" or "kip". Carnivore, darting after tiny crustaceans left by retreating waves. Northern populations sometimes winter on Long Island.
Habitat: Sandy beach, sandbars; water's edge

Least Sandpiper (*Calidris minutilla*)
Length: up to 6 in.   Migratory
Smallest shorebird. Darkish red-brown above; lighter brown chest with streaks; paler in winter; legs yellowish or greenish; bill thin, black, slightly downcurved. Call loud, shrill "kree". Carnivore, tiny crustaceans.
Habitat: Salt marsh, mudflats

Black Skimmer (*Rynchops niger*)
Length: up to 18 in.   Migratory
Named for habit of skimmimg water surface to attract and catch fish. Body black above, white below; short notched tail; big red black-tipped bill, lower mandible larger; red legs. Call abrupt bark. Carnivore, snaps up fish, often feeding at night.
Habitat: Barrier islands, lagoons, intertidal areas, shallow waters

Clapper
Rail

Least
Sandpiper

American
Redstart

Sanderling

Black
Skimmer

MTW 2001

Song Sparrow (*Melospiza melodia*)
Length: up to 7 in. Resident
Extremely common and variable perching bird. Brown back streaked darker brown; whitish underparts heavily streaked; grayish eyebrow; large dark spot in center of chest; legs and feet pinkish; long, rounded tail, pumped in flight. Call 3 short notes, then melodious trill. Omnivore, seeds, insects.
Habitat: Marsh, maritime forest, dunes, swale, thicket zone

Bank Swallow (*Riparia riparia*)
Length: up to 5 in. Migratory
Smallest swallow; nests in colonies, burrowing into cliffs and banks. Body dullish brown above, white below; distinct brown band across chest below white throat; notched tail; very rapid wingbeats. Call low, flat "chert, chert", buzzy notes or a twitter. Insectivore.
Habitat: Waterways

Mute Swan (*Cygnus olor*)
Length: up to 60 in. Resident
Introduced from Europe, now common. All white body, orange bill with prominent black knob at base; short, black legs and feet. Immature (cygnet) grayish-brown, gray bill. Neck held in "s" curve when swimming. Usually silent except for hiss. Herbivore, dipping over to feed. Aggressive.
Habitat: Coastal waters

Common Tern (*Sterna hirundo*)
Length: up to 15 in. Migratory
Familiar diving bird. Body silvery-gray above, whitish below; black cap and nape; bill reddish-orange with black tip; legs short, orange; tail deeply forked. In winter, forehead white, black shoulder bar. Call short "kip", long "kee-ar". Carnivore, plunges headfirst for fish, shrimp, insects.
Habitat: Beach, coastal waters

Least Tern (*Sterna antillarum*)
Length: up to 9 in. Migratory
Smallest tern. Body gray above, white below; black crown with white forehead and neck; bill yellow with black tip; legs short, yellow; tail short, forked. Call repeated "kip" or sharp "chee-eek". Carnivore, hovers over water then dives for minnows, shrimp, insects.
Habitat: Beach, coastal waters

Song
Sparrow

Bank
Swallow

Mute
Swan

Common
Tern

Least
Tern

MTW 2001

Ruddy Turnstone (*Arenaria interpes*)
Length: up to 10 in.   Migratory
Winters along mid-Atlantic, transient remainder of year. Calico orange and black back and wings, black markings on face and breast (duller in winter); white belly; short, slightly upturned black bill; short orange legs. Call "tuk-e-tuk". Omnivore, invertebrates and tern eggs.
Habitat: Sandy and rocky beaches, intertidal areas

Yellow Warbler (*Dendroica petechia*)
Length: up to 5 in.   Migratory
Small songbird, frequently seen bobbing its tail. Body bright yellow, greenish tinge to back; rusty streaks on male breast and sides; prominent dark eyes; bill slender, black; yellow spots on shortish tail; female duller overall. Call melodious "sweet-sweet-sweet". Insectivore.
Habitat: Maritime forest, thicket zone

Willet (*Catoptrophorus semipalmatus*)
Length: up to 5 in.   Migratory
Large sandpiper. Plumage speckled grayish brown; long blue-gray legs, straight gray bill. Flight reveals striking black wings with white stripe. Call a loud "pill-will-willet". Omnivore, invertebrates and fish.
Habitat: Salt marsh, mudflats

Marsh Wren (*Cistothorus palustris*)
Length: up to 5 in.   Migratory
Songbird with cocked tail. Back brown with white streaks; breast white; sides rusty brown; white stripe over eye; slightly down-curved bill. Call rattle, often preceded by buzz. Insectivore.
Habitat: Brackish and freshwater marshes

Lesser Yellowlegs (*Tringa flavipes*)
Length: up to 11 in.   Migratory
Transient visitor in spring and fall. Body slim speckled brown; white belly; thin black bill; yellowish-orange legs. Call a "tew-tew". Wades in shallow water and runs about intertidal areas in search of food. Omnivore, fish, invertebrates, berries. **Greater Yellowlegs** (*Tringa melanoleuca*)similar but larger body, longer bill (not illustrated).
Habitat: Salt marsh, intertidal area

Yellow
Warbler

Marsh
Wren

Rudy
Turnstone

Lesser
Yellowlegs

Willet

MᵀW₂₀₀₁

MAMMALS - Warm-blooded, with hair. Bear young live and nurse them with milk from mammary glands.

Harbor Seal (*Phoca vitulina*)
Body length: up to 5 ft.
Often seen in Long Island waters during winter. Smallish, chunky, color variable from yellowish-gray with dark mottling to almost brownish-black. Wide face, short muzzle, large black eyes, no visible ears. Spends much time basking on beaches and rocky shores, but can dive to 300' and stay submerged for 30 minutes. Carnivore, mainly fish, crustaceans, mollusks. One pup in early summer, born on land, able to swim at birth.
Habitat: Ocean coastal waters, bays, estuaries

Fin Whale (*Balaenoptera physalus*)
Body Length: up to 80 ft.
Largest animal seen off Long Island; recognized by small crescent-shaped dorsal fin. Sleek, flat-headed, bluish-black above, white below. Only whale colored asymetrically, right side of lower jaw white, left side dark. Numerous longitudinal grooves on throat; "v"-shaped snout with single median ridge on top; paired blowholes. Up to 400 baleen plates each side of upper jaw, white or streaked with purplish gray; used to strain pelagic crustaceans, squid, krill. Spout columnar, 15'-20' high, expands to ellipse; accompanied by loud whistle.
Habitat: Ocean, inshore and offshore

Right Whale (*Eubalaena glacialis*)
Length: up to 60 ft.
Most endangered whale on Atlantic and Pacific coasts. Body rotund, frequently with barnacles, brown to black, mottled, white on chin and belly; no dorsal fin. Large wide head one-third of body length has bumps and paired blowhole; lower jaw highly arched; 220 - 260 eight foot long baleen plates on each side for feeding on plankton. Flukes wide, pointed. Slow-swimming, easily approachable, buoyant upon death, which made it the "right" whale to hunt for oil and corset stays.
Habitat: Ocean, Long Island Sound; deep water

Harbor
Seal

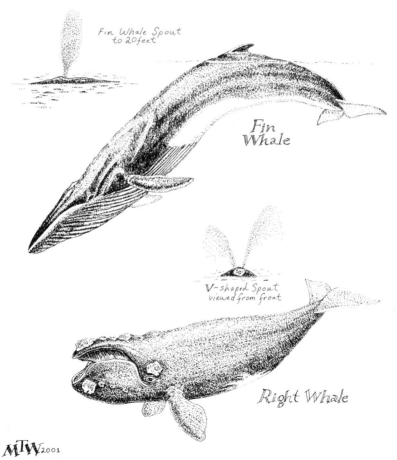

Fin Whale Spout
to 20 feet

Fin
Whale

V-shaped Spout
viewed from front

Right Whale

MTW 2001

GLOSSARY

Abdomen - Bottom or hindmost section of arthropod's body.

Accrete - Accumulate.

Achene - Small, hard, dry fruit with one seed.

Alternate - Occurrence of leaves on stem one after another, not opposite.

Ament - Long spike of unisexual flowers; catkin.

Anadromous - Migrating from ocean to breed in freshwater.

Angle of repose - The greatest stable slope of a deposit of sediment.

Annual - Plant that completes its life cycle in one growing season.

Aperture - The opening of a shell.

Aquifer - Porous layer of rock or sediment conducting groundwater.

Autotroph - Organism that synthesizes food from inorganic substances.

Axil - Upper angle where leaf joins the stem.

Baleen plates - Strips of bone, frayed along the edges, hanging from the roof of some whale mouths, used to strain food from the water.

Beach drift - Transportation of sediment by waves along a beach.

Beach face - The intertidal portion of the beach.

Benthic - Living on the bottom.

Berm - An accumulation of sediment formed by waves on the upper beach.

Bifurcate - To divide into two parts.

Bight - A broad embayment.

Biradial symmetry - Having both radial and lateral symmetry.

Blowhole - A nostril on the head of whales and other cetaceans.

Byssus - Bundle of threads used by some bivalves to attach to a surface.

Calcareous - Containing calcium carbonate, often chalky.

Callus - A thick calcareous growth over umbilicus of a gastropod's shell.

Canopy - Tallest vegetation of the forest.

Carapace - Hard, bony, outer covering; upper shell of turtle.

Carcinogen - A cancer-causing agent.

Cardinal teeth - In bivalves, interlocking calcareous projections at hinge.

Carnivore - Organism that eats only meat.

Carrion - Dead, decaying organism.

Cartilaginous - Having tough, elastic connective tissue.

Catkin - Dense, elongated cluster of scalelike unisexual flowers.

Cephalothorax - United head and thorax.

Chitinous - Covered by or composed of chitin, a horny substance.

Chondrophore - In some bivalves, a spoon-shaped depression on the hinge.

Cilia - Minute hairlike projections.

Columella - In gastropods, the solid or partially hollow axial pillar around which whorls are coiled.

Commensal - Association between two species, benefitting one, harmless to the other.

Community - Interacting group of various species in an area.

Continental shelf - The submerged sloping portion of the continent.

Crenulate - Scalloped, notched.

Crest - Elevated ridge between two wave troughs.

Decomposer - Plant or animal that breaks down organic matter.

Decumbent - In plants, main stem prostrate with ascending, erect branches.

Deposit feeder - Animal that ingests sediment to strain for food.

Detritus - Fragments of decaying plant or animal matter.

Dioecious - Having male and female flowers on separate plants.

Dorsal - Pertaining to the back.

Drupe - Fleshy fruit with single seed in hard covering (pit).

Ecosystem - Grouping of organisms within an environment.

Ecotone - Transitional zone between ecosystems.

Epibenthos - Freely moving bottom dwellers.

Epifauna - Animals on sediment surface attached to substrate.

Epiphyte - Plant that lives attached to another organism.

Exoskeleton - Hard supporting structure covering outside of body.

Fetch - Distance traveled by wind over open water.

Filter feeder - Animal that sucks in water to strain for food.

Floatables - Debris that floats on water.

Frond - Leaflike form in seaweeds.

Fusiform - Spindle-shaped swelling in central part, tapering at ends.

Glacial till - An unsorted mixture of rocks and finer sediment deposited by glacial ice.

Gonad - An organ that produces gametes.

Granulose - Composed of grains.

Halophyte - A plant that grows in saline soil.

Headland - A prominent scarp by the sea.

Herbivore - Animal that eats only plants.

Heterotroph - Organism which feeds on organic matter.

Holdfast - A rootlike attachment at the base of some seaweeds.

Hypoxia - Depleted oxygen levels harmful to organisms.

Indigenous - Native to a specific locality.

Infauna - Organism that lives in sand or mud.

Introduced - Originating from another country or region.

Lacustrine - Pertaining to a lake or pond.

Lag deposit - Heavier sediment that remains after lighter material has been removed.

Larva - Immature stage of an insect, between egg and pupa.

Lateral line - A midside sensory canal along the body of a fish.

Ligament - In bivalves, brown, chitinous attachment of the hinges.

Lirae - Threadlike ridges.

Littoral drift - The combined effect of beach and longshore drift.

Littoral transport - Littoral drift.

Loess - An unlayered deposit of windblown silt.

Longshore drift - Movement of sediment along the shore by currents.

Lunule - Usually heart-shaped impression on bivalve's shell.

Mantle - Outer layer of tissue that secretes the mollusk's shell.

Maxillary palp - Type of appendage around mouth in arthropods.

Medusa - Umbrella-like generation of coelentrates.

Moraine - A ridge of sediment deposited at the edge of a glacier.

Muscle scar - Indented, sometimes colored, attachment mark inside shell.

Neap tide - Tide with a minimum range at the time of first and third quarters of the lunar cycle.

Nekton - Free swimming animal.

Nematocysts - Stinging cells of coelenterates.

Niche - Specific role which an organism plays in its environment.

Nodule - Small lump.

Non-point source pollution - Contamination emanating from a diffuse source.

Nuchal - On turtle's carapace, scute closest to the neck.

Omnivore - Animal that eats anything edible.

Operculum - Plate or "door" covering the opening in a snail shell.

Opposite - Arrangement of leaves directly across from each other on stem.

Outwash plain - A flat area of sediments deposited by glacial meltwaters.

Ovate - Egg-shaped, with broadest part below middle.

Overwash - The surging of storm waters over the beach.

Panicle - Loosely branched, pyramidal flower cluster.

Parapods - Lateral extension of feet for propulsion.

Parietal - In gastropod shell aperture, upper portion of inner lip.

Pathogen - A disease-causing agent.

Pectoral fins - Thin, membranous appendages at shoulder.

Peduncle - Stalk or stem-like structure.

Pelagic - Living in open seas.

Perennial - Plant that lives three or more years.

Periostracum - Chitinous, skinlike outer covering on many mollusk shells.

Phyla - Broad categories into which plants and animals are divided.

Pincers - Jointed, claw-like grasping part.

Pistil - Female organ of a flower, containing ovules.

Pistillate - Having pistils but no stamens.

Point source pollution - Contamination emanating from a well-defined, localized source.

Polyp - Cylindrical body shape, attached at one end and mouth ringed with tentacles at other.

Prehensile - Adapted for holding or grasping.

Proboscis - A tubular sucking organ.

Protandric - Male when young, changing to female in later development.

Raceme - Elongated cluster of stalked flowers along a central stem.

Radial symmetry - Spoke-like arrangement of parts radiating from a central axis.

Radula - Rasping tongue of mollusks.

Receptacles - Fleshy structures in algae.

Resilium - Internal cartilage found in the ligament of some bivalves.

Rhizome - Horizontal rootlike plant stem that forms shoots above and roots below.

Salinity - Concentration of dissolved salts in water.

Scarp - Steep erosional slope.

Scavenger - Animal that feeds on decaying organic matter.

Scutes - Enlarged shieldlike scales on a reptile.

Sepals - Leaflike structures surrounding the petals of a flower.

Septum - Platform or partition.

Sessile - Attached directly at base.

Sheath - Tubular structure surrounding the lower portion of a leaf around the stem

Shoreface - The portion of the beach just below the intertidal zone.

Siphon - Tubelike body part for drawing or expelling water.

Siphonal canal - In gastropod shells, channel for siphon.

Spat - Immature bivalve, such as an oyster.

Spicules - Spikelike supporting structures.

Spire - Upper whorls of a snail shell.

Spit - A long narrow beach accreted into open water from a headland.

Spring tide - Tide with a maximum range at the time of new and full moons.

Stipule - Small, leaflike appendage at bottom of leafstalk.

Stolon - Lower branch that forms roots.

Subprovince - Biogeographical region characterized by species common throughout.

Substrate - Surface on which a plant or animal lives.

Succulent - Having fleshy stems or leaves.

Swash zone - Area of intertidal region where waves break.

Talons - Claws in birds of prey.

Telson - Last abdominal segment in crustaceans.

Tendril - Slender, coiling structure of climbing plants.

Tidal prism - Volume of freshwater exchanged with saltwater during a tidal cycle.

Tomentose - Covered with dense, matted hairs.

Trochophore - Ciliated larval stage of an oyster.

Trophic level - Classification based on feeding behavior of a particular group of organisms.

Trough - Depression between two wave crests.

Tsunami - Seismic sea wave.

Tussock - Mound of *Spartina* plants, peat, mussels and byssal threads.

Umbilicus - Hollow at base of snail shell.

Umbo - Prominent part of bivalve above hinge.

Understory - Shorter trees and shrubs of the forest.

Uropods - Short, immobile rod-like limbs on tail of some crustaceans.

Varices - Prominently raised vertical ridges on surface of gastropod shells.

Veliger - Advanced larval stage of an oyster.

Wave refraction - Redirection of waves as they enter shallow water.

Wrack - Organic and inorganic matter washed up on the beach.

Zooecium - Boxlike chamber in which a bryozoan resides.

Additional Sources of Information

Abbott, R. Tucker. 1968. *Seashells of North America*. New York: Golden Press.

Alden, Peter; Cassie, Brian; Kahl, Jonathan D. W.; Oches, Eric A.; Zirlin, Harry; and Zomlefer, Wendy B. 1999. *National Audubon Society Field Guide to the Mid-Atlantic States*. New York: Alfred A. Knopf.

Barton, Howard III and Pelkowski, Patricia A. 1999. *A Seasonal Guide to Bird Finding on Long Island*. Long Island, NY: ECSS: Sweetbriar Nature Center.

Bascom, Willard. 1980. *Waves and Beaches*. New York: Doubleday.

Lewis, Ralph S. and Stone, Janet Radway. 1991. "Late Quaternary Stratigraphy and Depositional History of the Long Island Sound Basin: Connecticut and New York." *Jounal of Coastal Research,* Special Issue, No. 11.

Long Island Shell Club. 1988. *Seashells of Long Island*. Long Island, NY: Long Island Shell Club, Inc.

Long Island Soundkeeper Fund. 1989. *The Sound Book*. Norwalk, CT: Long Island Soundkeeper Fund, Inc.

McCormick, Larry R.; Pilkey, Orrin H., Jr.; Neal, William J.; Pilkey, Orrin H., Sr. 1984. *Living with Long Island's South Shore*. Durham, NC: Duke University Press.

New York State Department of Environmental Conservation. 1994. *The Long Island Sound Study*. Albany, NY: New York State Department of Environmental Conservation Bureau of Publications.

Perry, Bill. 1985. *A Sierra Club Naturalist's Guide to The Middle Atlantic Coast - Cape Hatteras to Cape May*. San Francisco, CA: Sierra Club Books.

Rampino, Michael R. and Sanders, John E. 1981. "Evolution of the Barrier Islands of Southern Long Island, New York." *Sedimentology,* 28, 37-47.

Safina, Carl. 1997. *Song for the Blue Ocean: Encounters Along the World's Coasts and Beneath the Seas.* New York, NY: Henry Holt and Company.

Sirkin, Les. 1995. *Eastern Long Island Geology with Field Trips.* Watch Hill, RI: Book and Tackle Shop.

Sirkin, Les. 1996. *Western Long Island Geology with Field Trips.* Watch Hill, RI: Book and Tackle Shop.

Springer-Rushia, Linda and Stewart, Pamela G. 1996. *A Field Guide to Long Island's Woodlands.* Stony Brook, NY: Museum of L.I. Natural Sciences.

Stewart, Pamela G. and Springer-Rushia, Linda. 1998. *A Field Guide to Long Island's Wetlands.* Stony Brook, NY: Museum of L.I. Natural Sciences.

Stokes, Donald and Lillian. 1996. *Stokes Field Guide to Birds – Eastern Region.* United States: Little, Brown, and Company.

Teal, John and Mildred. 1969. *Life and Death of the Salt Marsh.* New York: Audubon/Ballantine Books.

Waldman, John and DiPaolo, Carol. 1998. *Hempstead Harbor: Its History, Ecology, and Environmental Challenges.* Sea Cliff, NY: The Coalition to Save Hempstead Harbor.

INDEX
(Illustration pages in bold print)

Field Notes

Field Notes